SpringerBriefs in Reproductive Biology

SpringerBriefs in Reproductive Biology is an exciting new series of concise publications of cutting-edge research and practical applications in Reproductive Biology. Reproductive Biology is the study of the reproductive system and sex organs. It is closely related to reproductive endocrinology and infertility. The series covers topics such as assisted reproductive technologies, fertility preservation, in vitro fertilization, reproductive hormones, and genetics, and features titles by the field's top researchers.

More information about this series at http://www.springer.com/series/11053

Ashok Agarwal · Luna Samanta
Ricardo P. Bertolla · Damayanthi Durairajanayagam
Paula Intasqui

Proteomics in Human Reproduction

Biomarkers for Millennials

 Springer

Ashok Agarwal
Cleveland Clinic
Cleveland, OH
USA

Luna Samanta
Department of Zoology
Ravenshaw University
Cuttack
India

Ricardo P. Bertolla
São Paulo Federal University
São Paulo
Brazil

Damayanthi Durairajanayagam
Faculty of Medicine
Universiti Teknologi MARA
Selangor
Malaysia

Paula Intasqui
São Paulo Federal University
São Paulo
Brazil

ISSN 2194-4253   ISSN 2194-4261   (electronic)
SpringerBriefs in Reproductive Biology
ISBN 978-3-319-48416-7   ISBN 978-3-319-48418-1   (eBook)
DOI 10.1007/978-3-319-48418-1

Library of Congress Control Number: 2016956832

Printed on acid-free paper

This Springer imprint is published by Springer Nature
The registered company is Springer International Publishing AG
The registered company address is: Gewerbestrasse 11, 6330 Cham, Switzerland

Foreword

Darwin's role in evolutionary biology is difficult to understand without the contributions of Gregor Mendel, James Watson and Francis Crick. Mendel found an outstanding pattern of inheritance associated with some explicit characters—the phenotype—that can be interpreted as the results of a series of heritable factors that are transmitted under simple mathematical rules from generation to generation. However, science had to wait until 1950s when J. Watson and F. Crick introduced the concept of molecular genetics. Since then, we have started to explore the connections between phenotypes (proteins), genes (basic information at the DNA/RNA) and genetic variability.

Thanks to the discovery of DNA and the modern technology, we now have a clearer understanding of the large variability in the DNA molecule, even within a single species. In fact, it is now recognized that it is not just the genes, but individual base arrangements along the genome in some cases which can cause certain diseases. Genomics was born for dealing with such phenomenon. However, the gene by itself is an inert unit. The protein, which originates from the information contained in the genome, is a critical effector of the phenotype. This is the reason why a large numbers of researchers nowadays are moving their focus on the final product of the genes, the proteins; and Proteomics was therefore born for such a purpose.

Reproductive life in the modern society is not following those strict rules to produce offspring as they were designed by nature. This is the reason why assisted reproductive techniques (ART) are able to help subjects that behave as "biologically sterile" to become an "assisted fertile". This SpringerBrief entitled "Proteomics in Human Reproduction: Biomarkers for Millennials" is an outstanding compilation summarizing the most relevant ideas that has been conducted to understand the relationship that we shall find between proteomics and male infertility, as well as how we may use the concept of proteomics in assisted reproduction. The book has been organized in a simple yet informative structure, dividing into 8 chapters, integrating the basic concepts of proteomics related to ART and identifying some of the doors that still need to be opened in this area with the currently available knowledge.

The contributors of this approach are affiliated to one of the most influential andrology centers in the world with a strong interest in male and female infertility. Dr. Ashok Agarwal, at the Cleveland Clinic, can be considered as an opinion leader in human reproduction related to male factors and the broad scope he has on how ART was and is evolving, is probably one of the reasons he has undertaken this new project. The soul of the book is visibly enthusiastic about the future roles of proteomics in assisted human reproduction targeting for more accurate and precise markers of infertility.

 Dr. Jaime Gosálvez, PhD graduated in Biology from the University of Granada (Spain, 1976), received his PhD from the University Autonoma de Madrid (1979) and today, he is the Chairperson and Professor in Genetics in the same institution. Jaime's research has been centred on reproductive biology for the last two decades. His current research is on translational biomedicine in the area of biology of reproduction. Specifically his interests include mammalian reproduction (male factor and sperm DNA damage assessment) with a special interest in human reproduction, endangered species reproduction and assisted reproduction management for industrial purposes. He has published over 350 indexed research contributions with 10 cover illustrations and has participated in 23 collective books and has been editor of two collective books. He has received maximum level of recognition according to Spanish standards for Research (Level 5 over 5) and Academia (Level 6 over 6) over a period of 35 years. He has been a principal investigator of 22 national and international competitive research projects. He is a supervisor of 15 PhD and 35 undergraduate research projects. He is also the inventor of 6 patents, all of which are currently being used. He is a scientific advisor of the Technological Park in Madrid and serves on the scientific advisory board of three different private companies in the area of biology of reproduction.

Department of Biology, Genetics Unit
Universidad Autónoma de Madrid, Spain   Jaime Gosalvez

Preface

The collective technologies that form the basis of the Omics revolution is changing the field of reproductive medicine and pushing research to greater heights. The use of the Omics technologies allows deeper examination into the proteomic basis (including the structure, function and interaction of the proteins involved) of a variety of specific infertility pathologies. Through the application of mass spectrometry and gel-based methods, proteomics investigations could potentially provide novel insights into the molecular basis of infertility. Moreover, the diagnostic and prognostic disease markers that could emerge from these studies may contribute towards the identification of possible therapeutic targets in the management of infertility. Proteomics in Human Reproduction: Biomarkers for Millennials provides an authoritative insight into the essential aspects of proteomics technologies, its role and application in male and female infertility as well as in assisted reproductive technology. Targeted at both clinicians and reproductive scientists involved in the field of human infertility, this brief aims to provide the reader with a thought-provoking and engaging review of the current findings and future possibilities of infertility-related proteomics studies. Moreover, the inevitable challenges that accompany the expanding area of proteomics are detailed along with a discussion on what lies ahead as proteomics research advances. This unique compilation of fundamental and comprehensive information pertaining to proteomics research associated with infertility will have an extensive appeal to both basic scientists in reproductive medicine and clinicians dealing with infertility. Furthermore, the authors are confident that this book will bridge the knowledge gap between laboratory-based researchers and clinical scientists who deal with patients seeking medical intervention for their fertility concerns. Ultimately via teamwork between the patient, clinician and researcher, the discovery of novel molecular biomarker(s) for infertility is poised to benefit subfertile couples. And with the refinement of high-throughput technologies, the goal of using biomarkers for the management of infertile couples could become a reality in the future.

Cleveland, OH, USA   Ashok Agarwal
Cuttack, India   Luna Samanta
São Paulo, Brazil   Ricardo P. Bertolla
Selangor, Malaysia   Damayanthi Durairajanayagam
São Paulo, Brazil/Cleveland, OH, USA   Paula Intasqui

Contents

About the Authors

Ashok Agarwal, Ph.D. is Professor at the Lerner College of Medicine, Case Western Reserve University and the Head of the Andrology Center and the Director of Research at the American Center for Reproductive Medicine, Cleveland Clinic Foundation, USA. He has researched extensively on oxidative stress and its implications on human fertility and has published over 550 original research articles and reviews. He serves on the editorial boards of several key journals in human reproduction and has edited over 32 medical text books/manuals related to male infertility, ART, fertility preservation, DNA damage and antioxidants. Ashok's current research interests include the study of molecular markers of oxidative stress, DNA fragmentation, and apoptosis using proteomics and bioinformatics tools.

Luna Samanta, Ph.D. is Professor in the Department of Zoology and the Dean of School of Life Sciences Ravenshaw University in India. Luna received a Raman Fellowship (December 2014–June 2015) from the Government of India for advanced research in proteomics of male infertility at the American Center for Reproductive Medicine, Cleveland Clinic. She was invited as a visiting scientist at Cleveland Clinic from June to July 2016. She has published 45 articles. Luna's research interests include systems biology approach to oxidative stress and sperm function using varicocele as a model.

Ricardo P. Bertolla, Ph.D. is Professor at the São Paulo Federal University in São Paulo, Brazil, and the Head of Research at the Urology Research Center of the São Paulo Federal University. Ricardo's research is on the cellular and molecular mechanisms of male infertility. The focus of his group is on sperm functional alterations and sperm and seminal plasma proteomics in different male infertility conditions, such as varicocele. Ricardo has published 42 articles. His current research is supported by the São Paulo Research Foundation, Brazil.

Damayanthi Durairajanayagam, Ph.D. is a Senior Lecturer in Physiology at the Faculty of Medicine, Universiti Teknologi MARA, Sungai Buloh Campus, Selangor, Malaysia. Damayanthi was a Fulbright Scholar, conducting advanced research in proteomics of male infertility at the American Center for Reproductive Medicine, Cleveland Clinic, USA, between 2012–2013. Her research interests include oxidative stress and the use of proteomics and bioinformatics in studying the molecular markers of oxidative stress in infertile males. She has published 17 articles.

Paula Intasqui, M.Sc. is a Ph.D. student at the São Paulo Federal University in São Paulo, Brazil, and was a visiting research student with a fellowship from the São Paulo Research Foundation, Brazil at the American Center for Reproductive Medicine, Cleveland Clinic, USA, from May to October 2016. Her work has been focused on sperm and seminal plasma proteomics associated with sperm function. Paula has 7 original research articles in the area of proteomics and male infertility.

List of Figures

List of Tables

Chapter 1
Introduction

Luna Samanta and Damayanthi Durairajanayagam

From the onset of civilization, human societies were concerned about the maintenance of their race. The International Committee for Monitoring Assisted
Reproductive Technology (ICMART) and the World Health Organization (WHO) in
2009 revised its glossary to define infertility as a disease of the reproductive system
where a couple is unable to achieve a clinical pregnancy after at least 12 months of
frequent sexual intercourse, without the use of any contraceptive method
(Zegers-Hochschild et al. 2009). As a result, assisted reproductive techniques
(ART) received a boost in the management of infertility and is continuously pro-
gressing to achieve new heights in addressing the problem with the constant evolution of genetics and molecular biology studies (Izzo et al. 2015).

1.1 Human Reproduction and the Current State of Infertility

Infertility remains as a significant disability worldwide and is estimated to affect
couples of reproductive age with a global incidence of about 9 % (Fig. 1.1) (Boivin
et al. 2007). While 20 % of the problem is attributed to male factor, a female factor
is reported in 38 % of cases. Anomalies in both men and women are observed in
27 % of cases and 15 % remain idiopathic without any attributable cause (de
Kretser 1997). In Southern and Central Asia, East and North Africa and Eastern

L. Samanta (✉)
Department of Zoology, Ravenshaw University, Cuttack, India
e-mail: lsamanta@ravenshawuniversity.ac.in

D. Durairajanayagam
Faculty of Medicine, Universiti Teknologi MARA, Sungai Buloh Campus,
Selangor, Malaysia
e-mail: damayanthi.d@gmail.com

© The Author(s) 2016
A. Agarwal et al., *Proteomics in Human Reproduction*,
SpringerBriefs in Reproductive Biology, DOI 10.1007/978-3-319-48418-1_1

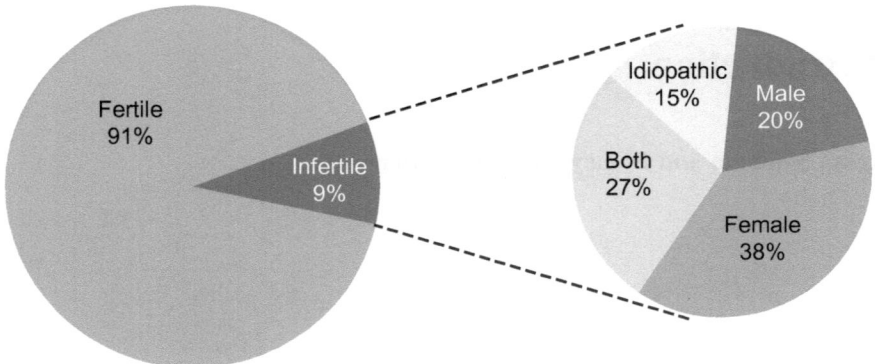

Fig. 1.1 Global prevalence of infertility

Europe, however, the incidence rate of infertility reaches approximately 30 %. For a young and healthy couple, the likelihood of achieving a pregnancy is about 20 % per menstrual cycle and approximately 90 % in a year of regular attempts, yet on an average in industrialized countries, approximately 10–15 % of couples are struggling to achieve a pregnancy (Boivin et al. 2007).

A plethora of factors are believed to be responsible for causing infertility in a normal couple and influencing the chance of getting pregnant naturally. The most important ones that warrant special mention are the woman's age, frequency of sexual intercourse, exercise, obesity, use of drugs or medications, smoking, infection, previous abdominal surgery, occupational and environmental exposure to toxicants, psychological factors, and stress (Izzo et al. 2015; Bretveld et al. 2007; Sharma et al. 2013d; Boivin et al. 2007). A systematic analysis of 277 demographic and reproductive health surveys between 1990 and 2010 has reported on the national, regional, and global prevalence of infertility in women from 190 countries (Mascarenhas et al. 2012).

The advent of ART in 1981 ignited a ray of hope for infertile couples. ART has played a role in 1 % of all babies born in the United States and 3 % in Australia per year. However, 70 % of the transferred embryos fail to implant (Coughlan et al. 2014). Another compounding problem of ART is multiple pregnancy that occurs when multiple embryos, rather than a single one, are transferred, thus exposing the patient to various complications. These include an increased risk of miscarriage(s), premature labor and birth; gestational hypertension, pre-eclampsia and diabetes in women pregnant with more than one fetus including ectopic pregnancy or a combination of normal and ectopic pregnancy (Gupta et al. 2015).

1.2   The Reproductive Time Line

The age of a man or woman is a pivotal factor that can affect fertility. Many couples are choosing to delay child-bearing in pursuit of education and career building. The reproductive time line in both men and women are naturally limited. Starting from puberty, human fertility peaks and then decreases over time. In men, testosterone biosynthesis starts declining as early as 35 years of age concomitant with a decrease in semen volume, sperm motility, and an increase in morphologically abnormal spermatozoa in the ejaculate (Stewart and Kim 2011; Dunson et al. 2004; Kimberly et al. 2012). Similarly, augmentation in sperm DNA damage with a corresponding decline in both motility and viability is observed after the age of 40 (Varshini et al. 2012).

The decline in male fertility is further strengthened by the fact that women with elderly partners have an increase in time to pregnancy (Mutsaerts et al. 2012). It is reported that partners of men over the age of 45, have a 4.6 times increase in time to pregnancy over 1 year that is elevated to 12.5 times over 2 years (Hassan and Killick 2003). Controlling for female age, infants born to fathers of $\geq 35–39$ years of age and older decreases exponentially compared to younger age groups (Matorras et al. 2011).

Women have a complex reproductive time line. A woman is born with a fixed number of oocytes and only 400–500 are actually ovulated during the reproductive age (Kimberly et al. 2012), thus limiting the woman's chances of becoming pregnant with advancing age. A woman under the age of 30 years has a 71 % chance of conceiving that declines to 41 % when she attains the age of 36 years or over (Mutsaerts et al. 2012), and the number of live births decreases exponentially after the age bracket of 35–39 years (Matorras et al. 2011).

Many factors are believed to be responsible for a woman to become pregnant and maintain a pregnancy, including oocyte's euploidy which is inversely correlated with the female age (Gianaroli et al. 2010). The rate of miscarriage's products aneuploidy for women over 35 years was reported to be 45.7 % versus 34.8 % for women under 35 years (Kroon et al. 2011). In comparison, Munné et al. reported the rates of euploidy in two consecutive assisted reproductive cycles of 50 % for women under the age of 35, 40 % for women between the ages of 35 and 39, and 33.3 % for women over the age of 40 (Munne et al. 2012). Furthermore, the risk of spontaneous abortion and implantation loss increases due to chromosomal abnormalities and aneuploidy with advancing age (Kimberly et al. 2012; Stewart and Kim 2011). Overall, women's fertility is significantly lower in their 30s and 40s (Kimberly et al. 2012).

Albeit the conventional tests, such as a routine semen analysis in males, or ovulation testing, ovarian reserve testing, and hysterosalpingography in females, including hormonal assays in both, that can help the clinician to develop the *modus operandi* for infertility treatment, yet the success rate of ART remains fairly low. Therefore, biomarker-based technologies are nowadays gaining importance for the fact that they allow for unbiased diagnosis of human infertility and informed clinical treatment decision-making. The rapid growth of molecular biology and laboratory technology has expanded to the point at which the application of technically advanced biomarkers will soon become even more feasible.

1.3   Biomarkers in Medicine

Biomarkers, defined as alterations in the constituents of tissues or body fluids, provide a dynamic and powerful approach to understand a spectrum of diseases with applications in observational and analytic epidemiology, randomized clinical trials, screening, diagnosis, and prognosis of a disease. More recently, the definition has been broadened to include biological characteristics that can be measured and evaluated as an objective indicator of either normal or pathological processes, or responses to therapeutic intervention (Biomarkers Definitions Working 2001). When put into practice, biomarkers encompass the means (tools and technologies) that can help to unravel the probability of a disease, its cause(s) or diagnosis, its progression or regression, or its treatment outcome (Naylor 2003; Etzioni et al. 2003; Mayeux 2004; Strimbu and Tavel 2010; Hulka 1990).

Over the years, various types of biomarkers have been used by scientists, physicians as well as epidemiologists to study human disease (Mayeux 2004). For example, blood pressure is used to determine the risk of stroke while serum cholesterol values are a biomarker and risk indicator for coronary and vascular disease. Thus, an ideal biomarker should be specific, easily detectable with minimum or no invasion, should be present at the earliest stages of the process under consideration, possess minimal associated side effects and be affordably priced (Kovac et al. 2013). The identification of novel biomarkers is a laborious and demanding process. Given that a biomarker may be a gene, protein, or messenger RNA, the possibilities are nearly endless. Once a biomarker has been identified and shown to be relevant, it must be brought into the clinical realm through validation. Validation refers to the need to characterize a biomarker's effectiveness or utility as a surrogate endpoint (Strimbu and Tavel 2010) (Fig.1.2).

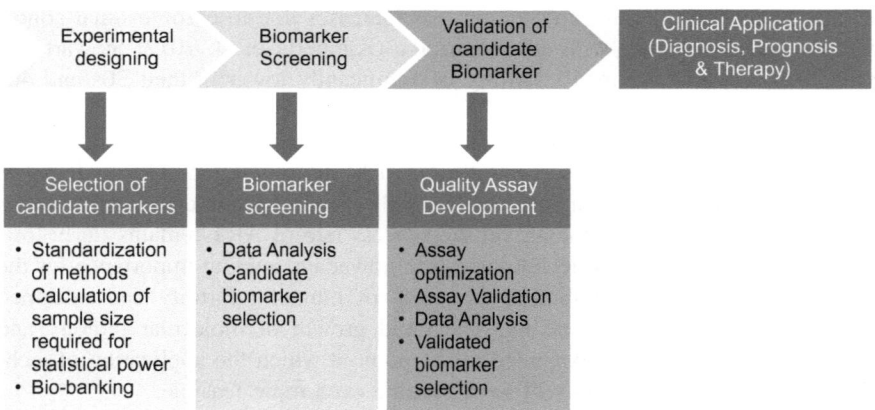

Fig. 1.2 Schematic representation of algorithm of biomarker development for clinical applications

1.4   Challenges in Fertility Evaluation and the Role of Biomarkers

An overriding goal of translational research is to define specific defective pathways in disease and to use this information to improve clinical diagnosis and treatment. Nevertheless, as is obvious in the work described so far for the field of human infertility (with a few exceptions), the translation from basic mechanisms to clinical care has been slow (Agarwal et al. 2014). However, important clues uncovered in animal models have begun to reveal a molecular understanding of the underpinnings of human infertility and thus to provide potential therapeutic or contraceptive targets to manipulate fertility pathways.

In many instances, despite the remarkable advances in the understanding of the complexity of sex determination, differentiation, gametogenesis, gamete function and fertilization in animal models, the reproductive defects seen daily in clinics are often not the focus of basic research. Most often, the diagnoses are either descriptive without explaining the mechanism or described as idiopathic. Nevertheless, advances in molecular and bioinformatics approaches in human reproductive biology and medicine have paved potential avenues for development of effective biomarkers to address fertility management.

1.5   Use of Omics to Discover Molecular Biomarkers

Spermatogenesis involves the formation of terminally differentiated haploid spermatozoa from diploid spermatogonia in the testis, while preserving the ability to contribute to a totipotent embryo which can effectively differentiate into a healthy individual. This complex process involves over 2300 genes, which are temporally and spatially regulated. This process not only depends on the genome, epigenome, and transcriptome, but also the proteome of the spermatogonia, supporting cells, and the generated spermatozoa (Carrell et al. 2016).

On the other hand, oogenesis and oocyte maturation involve a more complex phenomena where a vast number of intra- and extra-ovarian factors play pivotal roles. Oocyte meiotic and developmental competence is gained in a gradual and sequential manner during folliculogenesis and is related to the fact that the oocyte grows in interaction with its companion somatic cells. Huntriss' group observed the expression of a homeobox gene, Newborn Ovary Homeobox-encoding gene (NOBOX), from the primordial stage ovarian follicle through to the metaphase II (MII) oocyte (Huntriss et al. 2006). In addition, the dynamic expression profiles of 14 more homeobox genes throughout human oogenesis and early development was demonstrated. The study concluded that the homeobox gene transcripts in ovarian follicles and oocytes differ from those expressed in human blastocysts (HOXB4, CDX2, and HOXC9) and granulosa cells (HOXC9, HOXC8, HOXC6, HOXA7, HOXA5, and HOXA4) (Huntriss et al. 2006).

1.6   Proteomics as a Promising Tool for Biomarker Discovery

Proteomic strategies have been widely used in the field of reproductive biology for both basic and clinical research. Bioinformatics methods are indispensable for proteomic-based studies and help in understanding the biology of gametogenesis and aid in the discovery of potential biomarkers for the diagnosis and therapy of infertility.

In the post-genomic era, many proteome databases are developed that are available in public domain. For example, Gene Ontology (GO) terms annotate proteins into three general categories: biological process, molecular function, and cellular component. Similarly, other databases of biological processes such as Kyoto Encyclopedia of Genes and Genomes (KEGG), Reactome, PantherDB, protein domain analysis database (InterPro), protein–protein-interaction databases, such as IntAct, Biological General Repository for Interaction Datasets (BioGRID), Human Protein Reference Database (HPRD) and Search Tool for the Retrieval of Interacting Genes/Proteins (STRING), are among others that annotate proteins on the basis of pathway terms. Since single protein is often annotated with many terms from each database, protein–protein-interaction (PPI) network with statistical analysis is used to enrich its functional attributes (Agarwal et al. 2016a).

By applying bioinformatics annotation techniques, Zhu and co-workers have proposed an overview of spermatogenesis, from gene function to biological function and from biological function to clinical application. The authors demonstrated how by applying bioinformatics methods, the drug targets for sperm motility and scans for cancer-testis genes can be achieved from recently published proteomic data and studies. Thus, bioinformatics provides us with powerful connections between a list of genes and the hidden biological significance, thereby bridging between the basic research and clinical applications. With the development of post-genome projects, bioinformatics is helping scientists to interpret the integration of large datasets from proteomics studies (Zhu et al. 2013).

1.7   Proteomics in Human Reproduction Research and Its Clinical Application

Despite enhanced efficiency of ART, treatment of infertility is still far from being 100 % effective. One of the pivotal factors is the proper evaluation of germ cells, embryos and endometrium quality, in order to determine the actual likelihood to succeed. Presently the assessment system principally relies on morphological features of gametes and embryos. Although these strategies have improved the results, augmentation of the results warrants new diagnostic and therapeutic tools such as the Omics technologies (epigenomics, genomics, transcriptomics, proteomics, and metabolomics). These techniques offer a huge amount of information regarding the

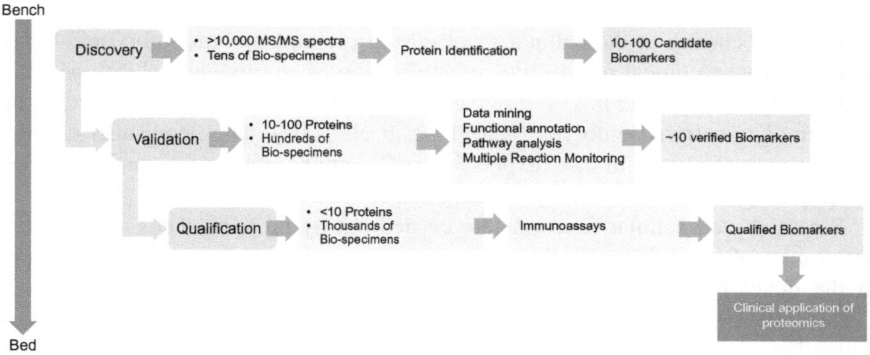

Fig. 1.3 Algorithm of proteomic biomarker discovery for clinical application

biological processes involved in reproductive success, thereby providing a broader view of complex biological systems with a relatively low cost and effort (Egea et al. 2014).

In this regard, clinical proteomics is an emerging field that seeks to apply this science in the search for biomarkers (Fig. 1.3), and the generation of protein profiles that can help predict, diagnose, and monitor human pathologies (Verrills 2006), such as infertility. In this case, informative protein profiles linked to optimal reproductive results can help in the improvement of diagnosis, fertility prediction, and the development of molecular strategies to select the best gametes and embryos and most receptive endometrium (Seli et al. 2007).

Although a wealth of literature on proteomic profiling during the last few years have enriched our understanding of reproductive medicine, there remains a paucity of information about cumulus and granulosa cells as well as oocyte due to the necessity of large numbers of oocytes to perform these analyses. Hence, most of the available information has been obtained from experimental animal models. Rolland et al. (2013) identified multiple potential biomarkers in reproductive tissues such as 83 proteins in testis, 42 in epididymis, 7 within seminal vesicles, and 17 in prostate. Their relevance is based on their participation in secretions present in seminal plasma, their effect on sperm quality, and their potential use as reproductive disorders markers.

Similarly, analysis of the follicular fluid proteome in women ≤ 32 years old revealed 11 potential protein markers that can predict ovarian response and live birth (Estes et al. 2009). An in-depth understanding of the embryonic proteome would serve as a true indication of cellular function and metabolism during mammalian preimplantation development. For instance, apoptotic and growth-inhibiting pathways are theoretical candidates to be closely involved in degeneration of embryos (Katz-Jaffe et al. 2006). These biomarkers provide a potential diagnostic platform for improving in vitro fertilization (IVF) procedures including in vitro culture conditions (supplementing media), stimulation protocols or cryopreservation techniques (Katz-Jaffe et al. 2005).

Identification of biomarkers will provide a mechanistic insight into the biological processes occurring at the cellular level during preimplantation embryonic development. From a clinical perspective, quantification of embryonic viability potential will result in an increase in IVF pregnancy rates and live births, while reducing the number of transferred embryos (Katz-Jaffe et al. 2006). This team also reported significant alterations in the expression of proteins related to morphological development of human blastocysts.

Essentially all cellular functions are carried out by proteins. Thus, comparative proteomics of normal versus diseased samples is deemed to be an important factor in the development of biomarkers for the diagnosis and treatment of diseases. Comparative proteomic analyses could possibly aid in the identification of biomarkers for non-invasive diagnosis of female diseases and assist in the prediction of success rates for ART. The promise of ART to achieve a pregnancy has led numerous infertile couples to opt for these procedures, with the hope of fulfilling their dream of having their own biological child. It is, however, noted that these techniques are not always associated with higher clinical pregnancy and live birth rates. To achieve a high pregnancy rate, it is important to understand the proteins involved in follicular microenvironment as well as the sperm proteins involved with fertilization and embryo development (Smith et al. 2007).

Chapter 2
Proteomics

Ricardo P. Bertolla

2.1 Introduction

Proteomics, the comprehensive identification and quantification of proteins present in a fluid or a cell population (James 1997), has become an important and established tool in understanding the mechanics of biological systems (Bantscheff et al. 2012), as proteins themselves are postgenomic effectors of cellular and molecular events (Zhang et al. 2013). Moreover, proteins and peptides may act as messengers, or interact upstream (controlling gene expression) or downstream in the Omics cascade, in order to regulate cell function or the extracellular environment (Zhang et al. 2013). In the case of complex samples, in which there is extensive crosstalk between different cell types and the environment—and the reproductive system is a great example of such a relationship—understanding the role these proteins may play, as well as potential regulators of their functions, is of great significance (Del Giudice et al. 2013; Amann 1989; Regassa et al. 2011).

Proteomics experiments may be of an exploratory or confirmatory nature. In exploratory (untargeted) proteomics, extensive mapping of digested peptides from a proteome is performed, in order to identify and quantify as many proteins as possible. This approach, coined shotgun proteomics (Claassen 2012), may be performed using chemical labeling of proteins or peptides in order to differentiate groups or may be label-free. A similar approach enriching postgenomic modifications may also be performed in order to observe post-translational protein modifications such as sumoylation, phosphorylation, oxidation, glycosylation, acetylation, and

R.P. Bertolla (✉)
São Paulo Federal University, São Paulo, Brazil
e-mail: rbertolla@yahoo.com

© The Author(s) 2016   9
A. Agarwal et al., *Proteomics in Human Reproduction*,
SpringerBriefs in Reproductive Biology, DOI 10.1007/978-3-319-48418-1_2

ubiquitination, among others (Olsen and Mann 2013). In confirmatory (targeted) proteomics, a number of techniques may be employed, ranging from conventional immunological labeling of differentially expressed proteins (i.e., Western blotting and enzyme-linked immunosorbent assay [ELISA]) (Bianchi et al. 2013; Ambekar et al. 2015; McReynolds et al. 2014) to mass spectrometry (MS)-based techniques (Lange et al. 2008).

Recent advances in MS techniques as well as in orthogonal protein and peptide separation technologies have dramatically increased throughput, sensitivity, dynamic range, and reproducibility of proteomics studies (Bantscheff et al. 2012). Coupled with advances in downstream data analysis, proteomic studies have impacted studies in the clinical sciences, with the promise of a personalized medicine tailored to an individual's disease (as opposed to the current generalized approach) (Cline et al. 2007; Hood and Flores 2012). This is especially true of the so-called bottom-up proteomics studies, in which proteins are broken down into peptides, and this complex mixture of peptides is separated by ultra-high pressure liquid chromatography (UPLC) and analyzed through tandem mass spectrometry (MS/MS) for sequencing of peptides against an in silico database constructed from sequenced genomes (Zhang et al. 2013).

However, it remains a major challenge of proteomics studies to translate findings complex in design, statistical analysis, and technical settings to answer clinical/biological questions. Bearing this in mind, this chapter is set up to discuss untargeted shotgun proteomics studies, including the usual steps in a proteomics experiment, ranging from sample preparation to separation techniques, MS equipment, and finally to the univariate and multivariate data analyses steps, as well as to functional enrichment tests usually performed for shotgun proteomics studies. Finally, we will also discuss results validation through a number of different techniques.

2.2 Methods Used in Proteomics

MS-based proteomics has profoundly impacted the field of proteomics in general, because of increased dynamic range, possibility to integrate with chromatographic separation equipment, increased sensitivity, and high quality identification of proteins (Bantscheff et al. 2012; Aebersold 2003; Aebersold and Mann 2003; Mallick and Kuster 2010). Mass spectrometers are able to detect the masses of ions [or the mass-to-charge ratios (m/z)] in a gas phase. In order to do so, mass spectrometers are generally subdivided into (i) an ionization source, (ii) mass analyzer(s), and (iii) a detector, of which the first two will be further discussed [for a review, see (Scherl 2015)].

Ionization of proteins and peptides was made possible with the development of the so-called soft ionization techniques; ionization techniques which lead to little or no fragmentation of the generated ion within the ion source, of which are of special note the matrix-assisted laser desorption/ionization (MALDI) (Tanaka et al. 1988) and electrospray ionization (ESI) (Fenn et al. 1989) techniques, both of which have been applied to proteomics studies. For shotgun proteomics studies, ESI allows for direct integration with liquid chromatography equipment coupled to tandem mass spectrometers (LC-MS/MS), which adds a quantitative aspect to the studies (Whitehouse et al. 1985).

Mass analyzers select, separate, or analyze ions based on their m/z values. Quadrupole (Q) mass analyzers, for example, select ions by varying an electric potential using radio frequency and direct current, focusing on ions of interest and diverting ions that are to be removed. Time-of-flight (TOF) analyzers base ion separation on the time it takes for an ion (of a specific mass and charge state) to fly through a vacuum tube. Ion traps are able to trap ions in a chamber and, by varying the applied electric field, allow for expulsion of single (or few) ions at a time. Finally, orbitrap (OT) mass analyzers detect m/z values by analyzing the cycle resonance of ions around a central electrode (Scherl 2015).

Current high-end mass spectrometers for shotgun proteomics studies utilize hybrid mass spectrometers, in which the first analyzer is a selector (such as a quadrupole or an ion trap), followed by a second analyzer which is converted into a collision chamber, followed by a high resolution mass analyzer, such as a TOF or OT. The collision chamber is used to generate fragments (fragment ions) of selected ions in the first analyzer (parent ions), and to determine the masses of these fragments, in an approach termed tandem mass spectrometry (MS/MS) (Scherl 2015). This is discussed in greater depth in the "*Identification of proteins using mass spectrometry*" section. A similar setup, utilizing a scanning quadrupole mass analyzer in the first step for separation of one or few parent ions and targeting specific fragments in the second analyzer may be used for specific protein detection/quantification, in selective reaction monitoring (SRM), multiple reaction monitoring (MRM), or parallel reaction monitoring (PRM) studies (Lange et al. 2008; Gallien et al. 2015). This will be explored further in the "*Validation of mass spectrometry data*" section.

In terms of setting up a general workflow for shotgun proteomics studies, the following steps are usually followed: (i) sample acquisition, storage, and preparation; (ii) protein and peptide labeling and separation; (iii) tandem mass spectrometry; (iv) database searching for protein identification/quantification; and (v) data analysis (Bantscheff et al. 2012). Enrichment of post-translational modifications may be performed, and a number of labeling approaches may be employed in order to increase sample normalization.

2.3 Sample Preparation

Upon sample collection, much care should be taken with regards to inhibiting proteases. This is especially important for analysis of post-translational modifications (PTMs), as many are subject to alterations due to protease activity (Olsen and Mann 2013). However, in the case of studies that are performed on ejaculated semen in humans, a special comment should be made with regards to semen coagulation/liquefaction. Coagulation occurs mainly due to semenogelin-1 and -2 polymerization. Prostatic Kallikrein-3 will then cleave these proteins at specific sites (at the N-side of Tyrosine residues), ultimately leading to semen liquefaction (Mitra et al. 2010). Thus, semen samples from humans have been subjected to protease activity (albeit specific and for a relatively short amount of time) even before the sample is available for separation of sperm/seminal plasma.

Protease inhibitors may be added individually to specific classes (such as metalloproteinase inhibitors), or commercially available protease inhibitor cocktails may be used (Clifton et al. 2011). Samples are then processed by separating cells (and cellular debris) from the extracellular fluid (seminal plasma, blood plasma, follicular fluid, etc.). This is usually achieved by an initial centrifugation at a lower force (not more than 600 xG) in order to avoid cell lysis and contamination of the supernatant fluid with cellular proteins. The supernatant is then centrifuged at maximum speed for a longer period of time (around 1 h) at cold temperatures (4 °C) to remove any remaining cellular debris. Next, the cell pellet can be washed a couple of times to remove any contaminating extracellular fluid, bearing in mind that lower centrifuge force is recommended to avoid cell lysis and loss of soluble proteins. The sample may then be kept frozen (at −70 °C or in liquid nitrogen) until sample preparation.

Sample preparation is different for cells compared to that for fluids. Cells must be lysed and proteins extracted. There are a number of different protocols for cell protein extraction, such as protocols for the extraction of membrane proteins, nuclear proteins, and whole cell lysate, among others (Tanca et al. 2013; Ahmed 2009; Weston et al. 2013; Intasqui et al. 2013a; Wisniewski et al. 2009). For fluids, most of the time protein extraction is not necessary, instead precipitation of proteins in order to remove contaminating lipids or sugars may be performed (Lo Turco et al. 2010; Camargo et al. 2013; Intasqui et al. 2013b).

The next step in the proteomics workflow is to consider options to decrease sample complexity. Current dynamic ranges of modern mass spectrometers utilized for shotgun proteomics are at or just above 4 orders of magnitude (according to their manufacturers), while protein dynamic range may achieve up to 9, or even 12 orders of magnitude (Anderson and Anderson 2002; Corthals et al. 2000). This means that highly concentrated proteins decrease the technical capability of detecting proteins at lower concentration. In seminal plasma, we have found that around 10 proteins represent 80 % of the proteome (*unpublished data*). In blood plasma (and its transudates, such as follicular fluid), albumin itself may represent 50 % of the proteome (Corthals et al. 2000). Thus, in order to detect (and quantify) proteins on the lower end, protein depletion and sample fractionating are two options that ensure a more comprehensive coverage of the proteome. Protein depletion is achieved by using columns which will bind to the desired protein (that is to be removed). There are several commercially available columns to remove albumin, immunoglobulins, and a number of other proteins (Zhang et al. 2013).

Protein fractionating may be performed in one-dimensional sodium dodecyl sulfate polyacrylamide (1D SDS-PAGE) gels, where they are usually separated by molecular mass, and strips of the gels containing a portion of the proteome are subjected to downstream analysis (Kim et al. 2003). Another option is pre-fractionating according to protein isolectric point—the pH at which their net charge is zero—(isoelectric focusing) in solid (Pernemalm and Lehtio 2013) or in liquid phase (Zuo and Speicher 2002). In either case, sample complexity is decreased, which leads to the possibility of observing proteins present in smaller amounts. Another orthogonal approach for protein fractionation is two-dimensional SDS-PAGE (2D SDS-PAGE), in which proteins are separated initially according to their isoelectric point in strip gels, and these are then submitted to gel electrophoresis for separation according to their molecular mass (Lopez 2007; Westermeier 2014).

2D SDS-PAGE experiments often target downstream MS protein identification to the differentially expressed gel spots (proteins). However, 2D SDS-PAGE is limited by the fact that (i) variability is quite high between different gels, and (ii) dynamic range is lower than that of LC-MS/MS platforms. Inter-gel variability has been dealt with by tagging proteins by fluorescent markers and then pooling samples from different groups to run in a same gel (Two-dimensional fluorescence difference gel electrophoresis—2D-DIGE), and dynamic range issues have been dealt with by depleting highly enriched proteins (Lopez 2007; Westermeier 2014). Also, an important advantage of 2D SDS-PAGE is that the initial mass of the protein spots is known before MS identification. Therefore, it is possible to know if the observed protein was intact or modified by cleavage, or by formation of doublets, for example. In seminal plasma, this may allow for visualization of specific semenogelin digests that present different biological roles (Robert and Gagnon 1999; Tomar et al. 2013; Mitra et al. 2010).

2.4 Identification of Proteins Using Mass Spectrometry

Because whole protein mass spectrometry is still limited, in terms of generating protein fragments for identification, protein samples need to be digested into peptides for identification and quantification (Bruce et al. 2013; Cox and Mann 2011). Trypsin is usually the enzyme of choice, because it produces peptides that fall into an optimal mass range (Vandermarliere et al. 2013). Trypsin cleaves proteins at the C-terminus of lysine and arginine residues, unless these are followed by a proline at the C-terminal end. Cleavage is highly specific and current protocols are quite efficient, leading to few missed cleavage sites (Vandermarliere et al. 2013). If a label-free approach is being employed, immediately following digestion, peptides are injected into the LC-MS/MS system (Bantscheff et al. 2012; Wisniewski et al. 2009; Cox and Mann 2011). If a labeling approach is chosen, peptides may be subjected to a number of different labels, such as dimethyl (multiplex peptide stable isotope dimethyl labeling) or isobaric tag for relative and absolute quantification (iTRAQ) (Bantscheff et al. 2012; Zhang et al. 2013; Wisniewski et al. 2009; Bruce et al. 2013; Cox and Mann 2011). This will be further discussed in the "*Quantitative proteomics*" section.

Proteins are inferred in LC-MS/MS experiments based on the mass of the peptides (MS parent ion, generated in the first mass analyzer of a hybrid mass spectrometer) and on the masses of detected fragment ions (MS/MS), usually reporting to specific amino acid sequences (detected in the second mass analyzer of a hybrid equipment) (Bantscheff et al. 2012; Aebersold and Mann 2003; Scherl 2015). In this data-dependent approach, an initial scan is performed in order to indicate MS peaks for fragmentation, followed by a step in which each peak of interest is filtered in the first mass analyzer, fragmented in a collision induced dissociation (CID) chamber, and its fragment ions detected in the second mass analyzer (Bantscheff et al. 2012; Scherl 2015; Bruce et al. 2013). The generated MS and MS/MS spectra are compared against in silico-generated mass values of putative proteins (based on sequenced genomes), leading to the identification of shared peptides and unequivocal peptides, the latter of which report to a single protein (Bantscheff et al. 2012; Cox and Mann 2011; UniProt 2015). A similar approach is used in data-independent experiments, in which the first mass analyzer is switched off, and the CID chamber alternates between high and low collision energies, generating parent and fragment ions with no need to filter out unwanted MS peaks (thus with loss of information) (Law and Lim 2013).

2.5 Quantitative Proteomics

Quantification of proteins in a sample, either in absolute terms or relative to another sample, is of importance if differential expression is to be studied. In order to achieve quantification, a few different approaches may be used, of which are of special note the label-free quantification (Washburn et al. 2001; Liu et al. 2004; Bantscheff et al. 2007), labeled quantification and absolute quantification methods.

In label-free experiments, quantitative information is extracted by (i) quantifying the total mass spectrometry signal for each peptide of a given protein (in other words, quantitation directly from constructed ion chromatograms generated during LC-MS/MS) or by (ii) determining the number of fragment spectra which report to peptides of a given protein (spectral counting) (Bantscheff et al. 2007, 2012; Washburn et al. 2001; Liu et al. 2004). Label-free shotgun proteomics has improved dramatically with the development of software packages specialized in aligning chromatograms from different runs and designed to normalize experiment-wise in terms of the amount of protein injected (Bantscheff et al. 2012; Zhang et al. 2013;

Cox and Mann 2008). Moreover, developments in chromatographic separation have greatly increased peptide resolution (in retention time), decreasing co-elution, and allowing for quantification of a large number of identified peptides (Ow et al. 2011; Delmotte et al. 2007).

In order to decrease intra-sample variability, labeled quantification experiments have been successfully devised. This may be possible through metabolic labeling or through chemical labeling. Stable isotope labeling with amino acids in cell culture (SILAC) is one of the best examples of metabolic labeling (Hoedt et al. 2014; Chahrour et al. 2015; Ong et al. 2002). In this type of experiment, labeled amino acids are fed to cells in in vitro culture conditions and, after a few rounds of division, it is assumed that translated proteins have incorporated these amino acids which, according to the label, will report to either one or the other condition. Thus, after protein extraction and digestion, samples may be mixed and analyzed in a single LC-MS/MS run, which in turn decreases inter-assay variation (Hoedt et al. 2014; Chahrour et al. 2015; Ong et al. 2002).

For studies on biological fluids or tissues not submitted to culture, chemical labeling of amino acids with isobaric markers, such as in iTRAQ, has allowed for simultaneous LC-MS/MS runs of up to eight different conditions (Chahrour et al. 2015; Choe et al. 2007; Ross et al. 2004). A similar, less costly, approach is the labeling of peptides using dimethyl labeling, which allows up to three-plex analysis in a simultaneous LC-MS/MS run (Boersema et al. 2009). A number of other labels are available, such as Tandem Mass Tags (TMT), Isotope-coded affinity tag (iCAT), among many more (Chahrour et al. 2015).

Finally, absolute quantification of proteins in shotgun proteomics studies have usually been performed by spiking into the sample previously digested peptides from one or more proteins of a different species, at a known concentration (Bruce et al. 2013). While this is not an error-free absolute quantification per se (absolute quantification in targeted proteomics requires spiking-in of peptides specific to the target protein), this allows for an internal standard normalization and a stoichiometric calculation for the studied proteome (Bruce et al. 2013).

2.6   Differentially Expressed Proteins

Regardless of the method of protein quantification, statistical analysis of the generated proteome should be performed under careful considerations. Initially, our group has utilized both a univariate and a multivariate approach to complex samples (Intasqui et al. 2013a, b, 2015, 2016; Camargo et al. 2013; da Silva et al. 2013; Lo Turco et al. 2013; Antoniassi et al. 2016; Del Giudice et al. 2016). In a univariate approach, statistical differences may be performed using standard parametric tests (Student's T-test, One-way ANOVA, etc.), provided that the data are robust (enough biological replicates, standard statistical assumptions for normality and homoscedasticity, which refers to homogeneity of variance). If not, the use of non-parametric tests (Mann-Whitney, Kruskal-Wallis, etc.) is a valid alternative.

Moreover, setting a minimal fold-change value may be desired, in order to filter out noise differences. Some studies report minimal fold-change values as low as 2 (or 0.5), although other studies have demonstrated that quantitative data utilizing spectral counting has been able to detect fold-change differences as low as 1.4. Other quality control assumptions may be used, such as quantification results derived from at least two different peptides. A complete list of the Minimum Information about a Proteomics Experiment (MIAPE) Mass Spectrometry Quantification (MIAPE-Quant) is under constant development and update by the Human Proteome Organization Proteomics Standards Initiative (HUPO-PSI) (Martinez-Bartolome et al. 2013).

Finally, in a multivariate data analysis approach, extraction of components using a Partial Least Squares Discriminant Analysis (PLS-DA) has been performed in the study of the seminal plasma proteome. These components are then subject to a binary logistic calculation which ultimately calculates odds ratio for group detection for each protein, as well as builds a multiplex data model to calculate group distribution according to protein quantification levels. This is particularly useful to suggest diagnostic models, although downstream absolute quantification studies will be necessary (Intasqui et al. 2015, 2016).

2.7   Proteomic Identification of Post-translational Modifications (PTMs)

One of the main mechanisms of biological variability is the wide array of PTMs present in proteins (Zhang et al. 2013; Olsen and Mann 2013; Clamp et al. 2007). Proteomic identification of PTMs—unbiased identification of multiple modifications in multiple proteins of a single cell type or fluid—presents additional challenges, because most PTMs are present in relatively low amounts, which leads to the need to enrich these modifications (Olsen and Mann 2013). Other than an enrichment step, a step to avoid degradation or artificial creation of PTMs, most of the proteomics workflow for the study of PTMs remains similar to a conventional proteomics workflow (Olsen and Mann 2013).

MS-based proteomics is the standard approach to identify PTMs in a proteomics workflow. However, enrichment of PTMs remains an important challenge, because, while for some modifications, such as phosphorylation and glycosylation, a relatively high efficiency is possible, other modifications usually require a larger amount of protein and protocol development (Olsen and Mann 2013). However, the promise of identifying multiple PTMs as associated to a specific biological condition is quite important, as it confers a functional aspect to a shotgun proteomics study. Indeed, differential protein expression does not shed light on all aspects of biological variability, while PTMs add an important functional modulation information to the proteome (Olsen and Mann 2013).

2.8   Interpretation of Protein Profiles Using Bioinformatics

Current proteomics studies are able to identify and quantify a high number of proteins—up to 5000 or 10,000 proteins have been identified from proteomes (Nagaraj et al. 2011; Beck et al. 2011). In seminal plasma, at least 2600 proteins have been identified with high confidence MS studies, over at least 4 orders of magnitude (da Silva et al. 2016). While this leads to a comprehensive biological view, it adds the need to filter out results which do not explain the biological question under study. In silico-derived protein–protein interaction (PPI) networks, functional enrichment of gene ontology terms and biological pathways in differentially expressed clusters, and multivariate statistical analysis are important approaches which may aid in filtering out a true signal from proteomic noise (Cline et al. 2007; Intasqui et al. 2013a, b; Camargo et al. 2013; Bindea et al. 2009).

In silico-derived protein–protein interaction networks are constructed based on large PPI databases, such as IntAct (Orchard et al. 2014), BioGRID (Stark et al. 2006), HPRD (Keshava Prasad et al. 2009), and STRING (Szklarczyk et al. 2015), among many others. Several software suites, such as MetaCore™ (Thomson Reuters, New York, NY, USA), Ingenuity Pathway Analysis® (IPA, QUIAGEN, Redwood City, CA, USA), and the open source platform Cytoscape (Cline et al. 2007), have made it possible to overlay identified proteins in a study to these interactomes, in order to suggest PPI subnetworks in any given study. Our suggested workflow is to (i) upload the list of identified proteins, (ii) identify PPIs between identified proteins, (iii) determine clusters specific/overexpressed in each group, and (iv) submit these clusters to functional enrichment studies. We have previously performed this workflow for the study of follicular fluid (Lo Turco et al. 2010), in seminal plasma (Intasqui et al. 2013b; Camargo et al. 2013; da Silva et al. 2013), and in sperm (Intasqui et al. 2013a) under different conditions.

For functional enrichment studies, functional annotations should be considered. The Gene Ontology (GO) Consortium has constructed a functional annotation database for a number of different species. Proteins, as gene products, inherit these functional annotations, which for the GO Consortium is subdivided into three major groups: Cellular Component, Biological Process, and Molecular Function. Each protein may receive a number of GO annotation terms, and these are shared by multiple proteins (Ashburner et al. 2000). If a cluster presents a higher frequency of GO terms than the whole GO database (or than a reference database, or when compared to the other group), these terms are said to be functionally enriched (Bindea et al. 2009; Maere et al. 2005). Many other ontology databases may be used, such as PantherDB (Mi et al. 2013), and highly curated databases exclusive to software packages, such as in MetaCore™ and Ingenuity Pathway Analysis®. Also, proteins may be annotated with pathway information, from such databases as Reactome (Croft et al. 2014), Kyoto Encyclopedia of Genes and Genomes (KEGG) (Kanehisa and Goto 2000), and PantherDB (Mi et al. 2013), among others. These pathways may indicate upstream and downstream interactions between the studied proteins and other molecules, such as lipids, sugars, and metabolites. Moreover, metabolic activity of the cell, and its assumed pathways brings information on response to the environment as well as on cell properties, such as senescence, aging, cell cycle activity, protein turnover (Merelli et al. 2014). This is particularly interesting when integrating metabolomics/lipidomics/steroidomics studies into proteomics studies. The Metscape app within Cytoscape, for example, utilizes pathway databases to generate a protein/metabolite/drug network with experimental data (Gao et al. 2010). Integration of these pathways may demonstrate not only the differentially expressed protein(s), but also its effects (such as metabolites) and effectors (such as steroid hormones) (Rouillard et al. 2015).

2.9   Validation of Mass Spectrometry Data

Validation of proteomics data usually falls into two major categories: validation through immuno-detection (Brown et al. 2013; Rifai et al. 2006; Surinova et al. 2011; Kohler and Seitz 2012; Kingsmore 2006) and validation using targeted mass spectrometry techniques (Lange et al. 2008; Gallien et al. 2015). Upstream validation (mRNA expression levels) and downstream validation (measurement of metabolites involved in the enriched pathways) are also important to consider as the former demonstrate the protein production cascade involved in the differential proteome, and the latter demonstrates that the assumed proteomic pathway is indeed active (Cline et al. 2007; Hood and Flores 2012).

Immunodetection may be performed on extracted proteins from the same sample or from a confirmatory prospective cohort, or may be performed directly on the studied cells, using fluorescent microscopy or flow cytometry. Of the many immune-detection methods available for extracted proteins, conventional ELISA are of special note (Brown et al. 2013; Rifai et al. 2006; Surinova et al. 2011; Kohler and Seitz 2012). In Western blotting, proteins are separated in 1D SDS-PAGE gels, transferred onto membranes which bind the proteins, and detected by the use of antibodies against the protein of interest. These intensities are usually normalized to constitutive proteins in the samples, in order to achieve lower intra- and inter-assay variation. ELISA, on the other hand, detects the protein(s) of interest by direct binding in microplates pretreated with antibodies against the protein of interest, signaling the presence, and intensity of the target protein (Scherl 2015).

Multiplexing immune-detection techniques are also currently available, utilizing the Luminex® system, which is an interesting development to consider for proteomics results validation, as they allow for the simultaneous intensity-based quantification of up to 50 different proteins (Purohit et al. 2015). However, while the available kits cover important pathways (such as oxidative stress), customization of detection kits for specific proteins is still prohibitive in terms of cost.

For detection of proteins directly on cells, two important techniques are immunohistochemistry (for tissue sections) or immunocytochemistry (for cells in suspension). Using fluorescently labeled antibodies allows for detection of populations marked or unmarked for the target protein (Erikson et al. 2007a, b). This is an interesting approach because cell integrity tests, such as membrane integrity, DNA fragmentation, or reactive oxygen species activity, may often be performed on the same sample, adding cellular biology information to proteomics results.

Mass spectrometry studies utilizing selected reaction monitoring (SRM), multiple reaction monitoring (MRM), and parallel reaction monitoring (PRM) are very sensitive (able to sense up to attomolar (10^{-18} mol/L) concentrations), specific, and allow for absolute protein quantification (Lange et al. 2008; Gallien et al. 2015). A specific workflow needs to be designed for each protein of interest, however, and in practice, this can lead to a long discovery pipeline (Bantscheff et al. 2012). In this type of experiment, a specific peptide from the target protein is filtered in a first analyzer of a mass spectrometer, and a (or a few) specific fragment(s) is detected on the second analyzer, leading to very high confidence quantification of that specific protein. In this experiment, calibration curves utilizing different quantities of the purified (or synthesized) target peptide (of the target protein) allow for determination of absolute quantities in each sample (Lange et al. 2008; Gallien et al. 2015).

Finally, it should be mentioned that current shotgun proteomics techniques utilizing high-end mass spectrometers produce results which, in themselves, are high confidence. While quantitative information from these experiments is subject to a number of variation factors, often coefficients of variation (CVs) from these studies fall between 20 % and 30 % (Piehowski et al. 2013). However, it is important to document specific conditions in which identification and quantification calculations were made, in order to guarantee that results are indeed high confidence (Martinez-Bartolome et al. 2013). In lacking a true validation, many studies report peptide mass errors, minimum number of peptides identified/protein, b- and y-sequencing results and false discovery rates, and this information allows for determination of MS conditions which, if stringent, would render reproducible results (Martinez-Bartolome et al. 2013, 2014).

2.10 Final Remarks

Proteomics is a promising and exciting area of research for integration of systems biology, which is especially applicable to complex samples and for integration into studies focused on the Omics cascade, ranging from mutations and polymorphisms through gene expression, protein expression, and down to metabolite generation and control of epigenetic factors. These systems biology studies aim at allowing for a more personalized medicine, and stem from the concept of phenotypic singularities due to genotypic potential within a given environment. The number of potential environmental agents is immeasurable, and individual response is potentially as widespread. Therefore, focusing on observing specific molecular fingerprints which associate to a biological condition my help to characterize disease. Generation of single biomarkers for diagnosis is one of the possible results, but most likely these studies will determine panels of differentially expressed proteins that associate to a phenotype, in that different nodes (proteins and other biomolecules) will weigh the biological network away or closer to a healthy state. In the case of infertility, which is often not a binary disease, the intensity of deviation from a healthy state of each individual in a couple will likely allow for understanding the causal factors of infertility and, potentially, therapeutic targets for intervention.

Moreover, while maintaining a Cartesian characteristic (in that it is hypothesis-driven), shotgun proteomics generates multiple hypotheses for downstream studies. On the other hand, much care should be taken, as is usually the case with technology-driven research platforms, in order to adhere to standards of quality that are under constant improvement. In the case of shotgun proteomics studies, the large amount of generated data is difficult to manually curate, which renders quality control of data acquisition fundamental, in order to increase repeatability. Data analysis and interpretation is also essential in order to transform lists of proteins into intra-assay variability is still quite high, due to technical limitations in peptide separation and identification. Targeted proteomics is, therefore, crucial for translation of results into actionable targets. It may well be that these results will in the future achieve the true nature of translational medicine in which the patient will benefit from novel diagnostic tests and potential druggable targets.

Chapter 3
Proteomics and Male Infertility

Nirlipta Swain, Gayatri Mohanty, Luna Samanta and Paula Intasqui

Several plausible factors are known to be associated with a reduction of the male reproductive potential, which may be congenital or acquired. Most of the anomalies underlying male infertility include urogenital abnormalities, varicocele, genetic abnormalities, endocrine disturbances, testicular failure, immunologic problems, cancer, systemic diseases, and infections of the genital tract (Tahmasbpour et al. 2014). Additionally, an altered lifestyle and exposure to gonadotoxic factors would further influence fertility of men (Barazani et al. 2014). In fact, a continuous decline in male fertility has been globally recorded by retrospective studies over the past few decades (Carlsen et al. 1992; Auger et al. 1995; Irvine et al. 1996; Geoffroy-Siraudin et al. 2012; Haimov-Kochman et al. 2012). However, the exact factors and mechanisms contributing to a large proportion of male infertility cases are still unknown, with such conditions being regarded as idiopathic.

The laboratory evaluation of male infertility remains an important area of research. Currently, initial diagnosis and evaluation of male infertility begins with the standard semen analysis in laboratories and hospitals. This includes assessment of semen volume, color, pH, liquefaction time, viscosity, sperm count and motility, sperm morphology, concentration of round cells and polymorphonucleocytes,

N. Swain (✉)
Department of Zoology, Ravenshaw University, Cuttack, India
e-mail: nirliptaswain@gmail.com

G. Mohanty
Department of Zoology, Ravenshaw University, Cuttack, India
e-mail: gayatri_mohanty32@yahoo.com

L. Samanta
Department of Zoology, Ravenshaw University, Cuttack, India
e-mail: lsamanta@ravenshawuniversity.ac.in

P. Intasqui
São Paulo Federal University, São Paulo, Brazil
e-mail: paula.intasqui@gmail.com

© The Author(s) 2016   21
A. Agarwal et al., *Proteomics in Human Reproduction*,
SpringerBriefs in Reproductive Biology, DOI 10.1007/978-3-319-48418-1_3

sperm agglutination and sperm viability (if required), based on the World Health Organization (WHO) Laboratory Manual for the Examination and Processing of Human Semen (World Health Organization, 2010). Nevertheless, its predictive value to assess sperm quality and fertility outcomes is limited. Moreover, men with a normal semen analysis report, normal history, and physical examination can also be infertile. Therefore, research related to sperm function and male fertility must establish more accurate diagnostic and prognostic methods for the management of male infertility and/or subfertility. With regards to this, routine evaluation of human semen is supplemented with several advanced tests, such as assessment of reactive oxygen species (ROS) levels, total antioxidant capacity (TAC), and sperm DNA fragmentation levels, DNA compaction and apoptosis, as well as genetic testing and evaluation of presence and localization of antisperm antibodies (Kashou et al. 2013; Kovac et al. 2013; Agarwal et al. 2015b; Homa et al. 2015; Sharma et al. 2016). Furthermore, the efficacy of standard semen analysis is further confounded due to the wide biological variability that can be present from sample to sample within the same individual (Bungum 2012). Therefore, the unraveling of the molecular factors that regulate key events in the process of fertilization holds great potential in our understanding of the causes of male infertility.

Advanced technologies like proteomics can become an efficient tool to investigate male infertility at a molecular level. The present chapter reports different infertility-related proteins identified in sperm, as well as seminal plasma and absent or with abnormal expression, which may provide the etiology of infertility in males. In acknowledging the contribution and involvement of these proteins in sperm physiology, the monitoring of the protein expression profile in infertile males could be used for identifying diagnostic biomarkers and as therapeutic targets.

3.1 The Journey of the Human Sperm: From Formation to Fertilization

Unlike females, who have a fixed reservoir of eggs, there is a continuous formation of spermatozoa in males. Spermatogenesis represents a complex succession of events, where the diploid and spherical spermatogonia undergo cellular divisions, and differentiation to produce haploid and elongated spermatozoa. Most of the stages of spermatogenesis occur in the seminiferous tubules of the testicles. Such a complex event requires precise expression of several proteins, whereas in mature sperm protein activity is regulated by highly efficient post-translational modification (PTM) machinery as the protein translation gets suspended.

The epididymis, a tortuously coiled structure atop the testis, serves many functions in male reproductive physiology. Although morphologically complete, yet the sperm released from the testis lacks the ability of motility and fertilization. During its transit through the epididymis, sperms are known to interact with several proteins secreted from the epididymal epithelium that are incorporated on the sperm plasma membrane. Active reabsorption of the sperm residual cytoplasm, changes in intracellular pH and ion concentration, as well as sperm chromatin remodeling and achievement of motility are some of the key changes that take place during the epididymal transit (Dacheux and Dacheux 2014; Cornwall 2009; Aitken et al. 2007). With the sperm losing its biosynthetic activity, the post-gonadal sperm differentiation in epididymis therefore, depends upon various physiological and biochemical transformations which includes the loss, modification, and/or remodeling of existing sperm proteins in response to cues delivered by the epididymis. Apart from aiding in sperm maturation, the epididymis also serves as a sperm reservoir and when ejaculation occurs, sperm is forcefully expelled from the tail end of the epididymis into the deferent duct.

As sperm transits through the male reproductive tract upon ejaculation, sperm are bathed in the seminal plasma secreted from male accessory sexual glands. The human seminal plasma is a mixture of secretions from the testis and epididymides (~ 10 %), seminal vesicles (~ 65 %), prostate gland (~ 25 %), and bulbourethral or Cowper's glands (~ 1 %) (Drabovich et al. 2014). Although primarily acting as a migrating fluid for the spermatozoa, the suspended proteins in seminal plasma play important roles in mediating sperm development, migration, and successful fertilization (Milardi et al. 2013; Drabovich et al. 2014). In maintaining a pH $\geq$ 7.2, the buffering properties of seminal plasma protects spermatozoa from the acidic environment of the vagina. Furthermore, the components of seminal plasma provide a nutritive *milieu* for motile spermatozoa, as well as help combat the detrimental effects of free radicals. Many seminal plasma proteins can attach to the surface of human spermatozoa and get involved in augmentation and inhibition of sperm motility, regulation of the immune response, interaction with the zona pellucida, modulation of the acrosome reaction, degradation of the extracellular matrix, and fusion with the oocyte membrane (Drabovich et al. 2014).

Upon leaving the epididymis, the sperm enter a convoluted series of tubules within the male genital tract that leads eventually to the penis. After ejaculation, the sperm travels beyond the vagina, into the cervix, uterine cavity, fallopian (oviduct) tube, and finally reaching the isthmus of the fallopian tube where fertilization usually occurs. Now, the fully matured spermatozoa are all set to take entry into the next phase of sexual reproduction, i.e., fertilization. The process of mammalian fertilization is a multifaceted and complex process as the site of semen deposition is far off from the site of fertilization in the female reproductive tract. Therefore, it is of utmost importance that the cellular processes that prime the sperm such as, capacitation, hyperactivation of motility, and acrosome reaction occur in a regulated fashion.

At the time of ejaculation into the female reproductive tract, the spermatozoa have the capacity to move. However, to gain its fertilizing ability, the sperm must initially undergo a process known as capacitation, which prepares the spermatozoa to bind to the zona pellucida (a glycoprotein matrix surrounding the oocyte) (de Lamirande and O'Flaherty 2012). Capacitation is basically a biochemical modification of sperm head membrane so as to prepare the sperm for acrosomal reaction and fertilization (Zaneveld et al. 1991). Although capacitation appears to be a continuous process that occurs throughout the sperm transit in the female reproductive tract in a stage-specific manner, hyperactivation seems to ensue more likely in the fallopian tube (Suarez and Pacey 2006). Being characterized as high amplitude flagellar movement changes, hyperactivated motility detaches the sperm from the endosalpingeal epithelium, and helps in penetration of oocyte vestments (Suarez 2008). Acrosomal reaction is almost the final step required for sperm–oocyte interaction. It is marked by exocytosis of acrosomal enzymes as the sperm makes its way through the zona pellucida, in order to degrade it to allow the sperm to fuse with the oocyte, forming the zygote (Brucker and Lipford 1995).

Regarding the sperm proteins, as the sperm travels through the female reproductive tract, the protein content of these cells is continuously altered, with some proteins being added to the sperm membrane, which prepares these cells for the tasks that they will have to perform, if successful fertilization is to result. Therefore, the journey of migration of spermatozoa from the site of insemination to the site of fertilization needs to be synchronized with several physiological changes in the female reproductive tract as well, that send the cues for acquisition of fertilization competence (Druart 2012).

Although several vital aspects of sperm maturation are known, many issues still remain pertaining to spermatogenesis and post-testicular sperm maturation that needs to be fully elucidated. It is also becoming increasingly apparent that these elaborate processes are carefully regulated by a myriad of gene products that are expressed and activated in a phase-specific manner (Hermo et al. 2010a, b). Therefore, unraveling the molecular factors such as the expression and activity of essential sperm and seminal plasma proteins that regulate the remarkable process of fertilization is extremely essential.

3.2 Spermatozoa as One of the Key Players of Reproduction—A Proteomic Approach

The complete knowledge of the structure and biochemical composition of sperm holds potential for a deeper understanding of these cells function at the cellular, molecular, and physiological level. In fact, spermatozoa are the only cells that fulfill its functions externally even when in a different individual, i.e., in the female genital tract, through a complex process of fertilization whereby the male genetic information in the form of the genome is transported to the oocyte.

Among the pioneer work on unraveling the proteome of human sperm was carried out by Naaby-Hansen et al. way back in 1990 (Naaby-Hansen et al. 1990). However, the application of liquid chromatography (LC) coupled to tandem mass spectrometry (LC-MS/MS) for in-depth proteomic analysis of human sperm was done much later in 2005 (Johnston et al. 2005). Till date, a large number of studies have been undertaken to prepare a protein profile of human sperm.

In 2006, Martinez-Heredia et al. utilized two-dimensional gel electrophoresis (2-DGE) to identify the sperm proteins in normozoospermic men. Over 1000 spots were observed in the reference gel, of which 145 spots were excised and identified by Matrix-Assisted Laser Desorption/Ionization-Time-of-flight mass spectrometry

(MALDI-TOF MS), allowing the identification of 98 proteins. These proteins were mostly involved with energy production and protein synthesis and folding (Martinez-Heredia et al. 2006). Another study was also set to put forth a high resolution 2-DE reference map of sperm proteins from human spermatozoa proteins from fertile sperm bank donors, with 3872 protein spots being identified (Li et al. 2007).

Moreover, studies were carried out to identify the origin of sperm proteins. Li and his group identified 112 proteins exclusively of testicular origin, 152 proteins exclusively of epididymal origin and 55 proteins common to both organs (Li et al. 2011a). They also reported that 47 % of the identified proteins were intrinsic sperm proteins expressed at the spermatid stage while 23 % were extrinsic sperm proteins that were acquired during epididymal transit and had an epididymal origin. Similarly, other studies reported 227 testis-specific proteins in mature human spermatozoa (Wang et al. 2013a), and 207 epididymal proteins present on spermatozoa (Li et al. 2010). In the later case, these proteins had highly specific locations on the sperm membrane and played specific roles on motility and protection against oxidative stress (Li et al. 2010). Furthermore, in order to unravel the capacitation-induced protein changes in ejaculated spermatozoa, Secciani et al. found that in capacitated sperm, there is a reduction of proteins related to protein fate, metabolism, and flagellar organization, while a marked increment was observed for proteins related to cellular stress (Secciani et al. 2009).

With spermatozoa being a highly differentiated cell, subcellular proteomic studies on sperm cells also gained greater interest in recent years. With the sperm nucleus being the major portion transmitted to oocyte, a list of 581 chromatin or nuclear proteins in human sperm cell was reported (Castillo et al. 2014). It was found that 56 % of sperm nuclear proteins may have an epigenetic activity, with functions such as chromosome and chromatin organization, protein-DNA complex assembly, DNA packaging, gene expression, transcription, chromatin modification, and histone modification. Previously, another study had identified 403 different proteins from the isolated human sperm nuclei (de Mateo et al. 2011). The same group has also, for the first time, described the correlation between proteomics, DNA integrity and protamine content (de Mateo et al. 2007). Baker et al. carried out the largest reported compartmentalized proteomic analysis of human spermatozoa, which showed that out of 1429 identified proteins, 721 proteins are exclusively found in the tail and 521 exclusively in the head (Baker et al. 2013). A study on human tail proteome revealed that there are two main categories of proteins in the sperm tail, namely, those related to sperm tail structure and motility with the majority being those involved in metabolism and energy production (Amaral et al. 2013). In addition to the above subcellular fraction studies, it is imperative to study the sperm surface proteome by using purified plasma membrane fractions. With this aim, a specialized research has targeted to prepare a comprehensive coverage of the proteins on the human sperm surface. In this study, it was demonstrated that sperm membrane presents many markers of membrane rafts, which are redistributed to the peri-acrosomal region after capacitation and are capable of binding to the oocyte zona pellucida. Therefore, these membrane rafts may play an important role during sperm-oocyte interaction. Additionally, several important proteins for fertilization were also identified (Nixon et al. 2011).

Recently, based on an extensive literature search, a compiled list of all the sperm proteins described to date and their potential functional implications was reported (Amaral et al. 2014a). About 6198 proteins were identified with different functional implications in sperm such as, metabolism, apoptosis, cell cycle, meiosis membrane trafficking, and even RNA metabolism and translational regulation.

Moreover, several proteomic approaches have been undertaken to detect the PTM protein profiling of spermatozoa. Phosphoproteomic studies have been extensively targeted since it plays a crucial role in the regulation of almost all sperm functions. Naaby-Hansen et al. carried out a preliminary study to identify the phosphorylation sites in sperm proteins (Naaby-Hansen et al. 1997). Recently, Wang et al. investigated the overall phosphorylation events in sperm and identified functional kinases during human sperm capacitation (Wang et al. 2015). A total of 3303 phosphorylated sites, corresponding to 986 phosphorylated proteins, were recognized using immobilized metal affinity chromatography or titanium dioxide beads (IMAC-TiO$_2$) phosphopeptide continuous enrichment methods by LC-MS/MS, of which the phosphorylation levels of 231 sites (including 25 tyrosine sites) were significantly increased. Even comparative phosphoproteomic reports associating the role of phosphorylation in asthenozoospermic patients have also been carried out (Chan et al. 2009; Parte et al. 2012). Similarly, the other PTM comprehensibly studied in sperm is the lysine–acetylation. Sun and his group characterized 1206 lysine-acetylated sites, corresponding to 576 lysine-acetylated proteins in human capacitated sperm (Sun et al. 2014).

In another study by Yu and his group, 973 lysine-acetylated sites were identified that matched to 456 human sperm proteins, including 671 novel lysine-acetylated sites and 205 novel lysine-acetylated protein (Yu et al. 2015). The other PTM critically analyzed by several researchers in sperm includes glycoproteome profiling by Wang et al. (2013b), S-nitrosoproteome profiling by Lefievre et al. (2007), histone methylome by Krejci et al. (2015), and proteomic characterization of different SUMOylated proteins in human ejaculated sperm by Vigodner et al. (2013). Not only it is apparent that proper PTM is a prerequisite for normal functioning of sperm, anomalies in the occurrence of PTMs in spermatozoa proteins may lead to its dysfunction resulting in infertility (Samanta et al. 2016).

3.3 Seminal Plasma as a Promising Source of Biomarkers of Male Infertility

Proteins of seminal plasma are relevant for sperm function and relate to sperm interactions with the various environments along both the male as well as the female genital tract. As the alterations at the molecular level in seminal plasma can affect male fertility, proteomic analysis of seminal plasma would be imperative as noninvasive clinical diagnostics of male reproductive system disorders. Targeting the dysfunctional proteins in infertile patients would facilitate the identification and management of these conditions through screening, early diagnosis, and more accurate prognosis. A comprehensive account of the seminal plasma protein repertoire of fertile males would help us to better understand the normal molecular composition of semen. On the other hand, a comparative proteomic analysis of seminal plasma from men with different spermatogenic impairment could be used to find specific male infertility marker proteins, given that 10 % of the seminal plasma proteome is originated in the testis and epididymis (Batruch et al. 2011).

A proteome map of normal human seminal fluid was presented by Fung et al., who identified over 100 protein and peptide components to be a part of normal human seminal plasma, such as semenogelin-1 (SEMG1), semenogelin-2 (SEMG2), prostate-specific antigen (KLK3), prostatic acid phosphatase (ACPP), albumin (ALB), and prolactin inducible protein (PIP) (Fung et al. 2004). A more comprehensive analysis was performed later by Pilch and Mann. Taking three semen samples prior to liquefaction from a single fertile male as a model system, they reported the high-confidence identification of 923 proteins in seminal plasma (Pilch and Mann 2006). The majority was cellular proteins, but a large number of extracellular or secreted proteins were also reported. This high-confidence characterization of seminal plasma content provides an inventory of proteins engaged in metabolic activities, mainly enzymatic and signal transduction, as well as some involved in immune responses. Proteins from the entire male reproductive system were observed in the seminal plasma, presenting 66 % of homology with the prostate fluid and 18 % with the epididymal fluid, while 38 % were common to all fluids (Fu-Jun and Xiao-Fang 2012). Another characterization of the seminal proteins in fertile men was performed by Milardi et al., and the seminal plasma from five proven fertile men was analyzed. They reported that the common proteins present in all samples included mainly SEMG1, SEMG2, olfactory receptor 5R1 (OR5R1), lactotransferrin (LTF), cathelicidin antimicrobial peptide (CAMP), spindlin-1 (SPIN1), and clusterin (CLU) (Milardi et al. 2012). The Gene Ontology (GO) annotation analysis showed that most of the proteins had binding activity, while catalytic activity, structural molecule activity, and enzyme regulation were also observed. In two separate studies, it was also demonstrated that seminal plasma proteome differs from that of other human body fluids, such as blood, saliva, and vaginal fluid (Yang et al. 2013; Orphanou et al. 2015).

To confirm if protein biomarkers of the male genital tract can be identified in the seminal plasma, Rolland et al. performed a proteomic analysis in non-liquefied seminal plasma from one healthy donor and combined the results with data from previous studies and with transcriptome profiles of different reproductive tissues (Rolland et al. 2013). Using this integrative approach, they identified a total of 2545 seminal plasma proteins, of which 83 proteins from the testis, 42 from the epididymis, 7 from the seminal vesicles and 17 from the prostate. In order to identify which seminal plasma proteins are originated from the testicular and epididymal fluids, the proteomes of pooled seminal plasma from fertile controls and post-vasectomy (PV) men were compared (Batruch et al. 2011). The group cataloged 32 proteins exclusively observed in the control group (testicular and epididymal proteins), and 49 underexpressed proteins in PV men (proteins expressed in the testis and epididymis, but also in other organs from the male reproductive tract). Some of the identified testicular and epididymal proteins were testis-expressed sequence 101 protein (TEX101), phosphoglycerate kinase 2 (PGK2), histone H2B type 1-A (HIST1H2BA) and glyceraldehyde-3-phosphate dehydrogenase, testis-specific (GAPDHS).

In another study, a classification of 331 seminal plasma proteins based on biological annotation revealed that about 44.4 % were proteases, 19.03 % were signal transduction proteins, 15.4 % were transport-related proteins, 11.5 % were enzyme regulators, 6.3 % were involved with programmed cell death, 3.62 % were structural proteins, and 3.32 % were cell movement-associated proteins, whereas 17.8 % had unknown molecular function (Bai et al. 2009).

Another study had aimed to carry out a functional characterization of heparin-binding proteins (HBPs) present in human seminal fluid, as these may be directly involved in sperm capacitation and acrosome reaction (Kumar et al. 2009). Functional analysis revealed that 38 % of HBPs presented enzymatic activity, 20 % were involved in RNA processing and transcription, 18 % were structural proteins or were involved with transport, and 16 % had important functions in cell recognition and signal transduction. Recently, utilizing a proteomic set up, a profiling of N-glycosylated proteins in seminal plasma was also carried out (Yang et al. 2015), which could be a resource for further screening of biomarkers for male diseases, including cancer and infertility at the level of N-glycosylation.

In addition to the characterization of the complete seminal plasma proteome, efforts have also been made to identify the protein content of microvesicles present in the seminal plasma. In epididymosomes, for instance, 146 different proteins were identified, which may be added to the sperm membrane to promote their maturation (Thimon et al. 2008). It was also demonstrated that the epididymosome protein content differs from that observed in prostasomes (Chiasserini et al. 2015).

3.4 Proteomics in Various Infertility Conditions and Their Possible Biomarkers

A great deal of evidence has suggested that the quality of human semen is deteriorating. With most of the current clinical fertility tests available in the market failing to detect as well as explain the etiology of male infertility condition, the need to authenticate newer and better investigative technologies for diagnosis is increasingly being recognized. A comparative proteomic profiling of fertile samples and samples collected from infertile males would reveal the anomalies in protein expression. These protein biomarkers could not only be used as diagnostic measures, but also as a guide to find better therapeutic solutions. Although some studies were performed to compare the seminal plasma and sperm proteomes between fertile and infertile men, without focusing on the cause of the infertility (Xu et al. 2012; Thacker et al. 2011; Cadavid et al. 2014), the majority of studies compared control men with infertile men with different infertility conditions, such as azoospermia and varicocele, which will be further described in the following sections and summarized in Table 3.1.

3.4.1 Azoospermia

The most extensively studied infertility condition so far is azoospermia, a term that refers to ejaculate that lack spermatozoa without implying a specific underlying cause (Esteves and Agarwal 2013). Azoospermia is diagnosed in 20 % of subfertile men and has two forms: obstructive azoospermia (OA) and non-obstructive azoospermia (NOA). The appropriateness of the term azoospermia and the reliability of diagnosing the absence of spermatozoa have been the focus of debate over the past decade. OA is caused by a physical obstruction in the male reproductive tract. The biological outcome of OA is thus identical to that of vasectomy, which is a surgical severance of the vas deferens. NOA is a more complicated infertility syndrome with the azoospermia being secondary to a failure to produce sperm. Thus, NOA may be further sub classified as maturation arrest, Sertoli cell-only syndrome, and hypospermatogenesis. The reliable diagnosis of the absence of spermatozoa in a semen sample is important for diagnosing male infertility, ascertaining the success of testicular sperm extraction (TESE), and determining the efficacy of hormonal contraception (Aziz 2013). For most men with azoospermia, testicular biopsy is the only currently used method to definitively distinguish between OA and NOA. Thus, there is an urgent need for an alternative noninvasive approach with better diagnostic potential. Moreover, in one recent study, the proteomic profiling of sperm indicated that OA patients might produce antibodies against two sperm proteins, tektin-2 (TEKT2) and triosephosphate isomerase (TPI1), which might be the causative factor (Zangbar et al. 2016).

With no biomarkers currently existing for the definitive differential diagnosis of OA and NOA, it is therefore plausible that some of these proteins of ejaculated seminal plasma may be useful as noninvasive biomarkers to discriminate NOA from OA. The first proteomic studies with this aim were performed by Starita-Geribaldi et al. (2001, 2003), and included fertile men, vasectomized men (PV, simulating OA), and men with Sertoli cell-only syndrome (NOA). In the first study, 757 spots were detected in fertile men, with 2 main spots identified as ACPP and KLK3. In NOA, eight spots were not observed in all patients when compared to fertile men. One spot was present in all fertile men and absent in all infertile men (Starita-Geribaldi et al. 2001). In the follow-up study, 937 different spots were observed, some of which were identified as CLU, zinc-alpha-2-glycoprotein (AZGP1) and glycodelin S (PAEP), in addition to ACPP and KLK3, already identified in their previous study. Five proteins were suggested as biomarkers of azoospermia, i.e. CLU (acidic train), serum amyloid P-component (APCS), human epididymal secretory protein E1 (NPC2), cysteine-rich secretory protein 1 (CRISP1), Cu/Zn superoxide dismutase (SOD1) (Starita-Geribaldi et al. 2003).

Table 3.1 Comparative proteome profile of human semen associated with infertility and the identification of probable biomarkers of infertility

Condition	Sample	Methods used	Proteins identified	Reference
Azoospermia	Seminal plasma	2D-MALDI-TOF MS	ACPP and KLK3	Starita-Geribaldi et al. (2001)
Azoospermia	Seminal plasma	2D-MALDI-TOF MS/MS	CLU, AZGP1, PAEP, APCS, NPC2, CRISP1 and SOD1	Starita-Geribaldi et al. (2003)
Azoospermia	Seminal plasma	2D-DIGE-LC-MS/MS	STAB2, CP135, GNRP, PIP, NPC2	Yamakawa et al. (2007)
Azoospermia	Seminal plasma	Strong cation exchange chromatography-LTQ-Orbitrap MS/MS	COL6A2, GGT7, SORD, PGK2, LDHC, ZPBP2 and ELSPBP1	Batruch et al. (2012)
Azoospermia	Seminal plasma	LC-MS/MS	LDHC, SPAG11B, MUC15, TEX101 and CEL	Drabovich et al. (2011)
Azoospermia	Seminal plasma	2-DIGE-MALDI-TOF-TOF MS/MS and 2-DIGE-LC-LTQ-Orbitrap MS/MS	PAP	Davalieva et al. (2012)
Azoospermia	Seminal plasma	SRM	TEX101 and ECM1	Drabovich et al. (2013)
Azoospermia	Seminal plasma	1D-NanoLC-MS/MS	LGALS3BP	Freour et al. (2013)
Azoospermia	Seminal plasma	Immunocapture-SRM	TEX101	Korbakis et al. (2015)
Azoospermia	Spermatozoa	2D-MALDI-TOF/TOF MS	TEKT2 and TPI1	Zangbar et al. (2016)
Varicocele	Spermatozoa	2D-DIGE	HSPA5, SOD1, ATP5D	Hosseinifar et al. (2014)
Varicocele	Spermatozoa	1D-LC-LTQ-Orbitrap MS/MS	CRISP2 and ARG2	Agarwal et al. (2015b)

(continued)

Table 3.1 (continued)

Condition	Sample	Methods used	Proteins identified	Reference
Varicocele	Spermatozoa	1D-LC-LTQ-Orbitrap MS/MS	GSTM3, SPANXB1, PARK7, PSMA8, DLD, SEMG1, and SEMG2	Agarwal et al. (2015c)
Varicocele	Spermatozoa	1D-LC-LTQ-Orbitrap MS/MS	TEKT3 and TCP11	Agarwal et al. (2016b)
Varicocele	Spermatozoa	1D-LC-LTQ-Orbitrap MS/MS	HSPA2, ODF2, CCT6B	Agarwal et al. (2016c)
Varicocele	Seminal plasma	NanoUPLC-ESI-MS[E]	G3P, PARK7, SOD, S100-A9, MDH	Camargo et al. (2013)
Varicocele	Seminal plasma	2D-ESI-QTOF MS/MS	UPP1, SEMG1, SEMG2, PIP, CYTS	Zylbersztejn et al. (2013)
Varicocele	Seminal plasma	2D-ESI-QTOF MS/MS	ALBU, CLU, SEMG1, SEMG2 and PSMA6	Del Giudice et al. (2013)
Varicocele	Seminal plasma	NanoUPLC-ESI-Orbitrap MS/MS	CAB45 and CRISP3	Del Giudice et al. (2016)
Asthenozoospermia	Spermatozoa	2D-PAGE	COX6B, HIST1H2BA and HSPA2	Martinez-Heredia et al. (2008)
Asthenozoospermia	Spermatozoa	2D-Nano-HPLC-ESI-MS/MS	PTPN14	Chao et al. (2011)
Asthenozoospermia	Spermatozoa	NanoUPLC-MS[E]	HSPs	Parte et al. (2012)
Asthenozoospermia	Spermatozoa	2D-MALDI-TOF-TOF MS	COX6B and HSPA2	Hashemitabar et al. (2015)
Asthenozoospermia	Spermatozoa	2D-MALDI-TOF MS	PATE1	Liu et al. (2015)
Asthenozoospermia	Seminal plasma	1D-LC-MS/MS	PARK7	Wang et al. (2009)
Globozoospermia	Spermatozoa	2D-MS	ZNF174 and CAPZA3	Luo et al. (2008)
Necrozoospermia	Spermatozoa	2D-DIGE-MALDI-TOF MS/MS	SPANXa/d, SAMP1 and ODF2	Liao et al. (2009)

(continued)

Table 3.1 (continued)

Condition	Sample	Methods used	Proteins identified	Reference
Oligoasthenozoospermia	Seminal plasma	LC-LTP-Orbitrap MS/MS	TBCB, AACT and ALDR	Herwig et al. (2013)
Oligoasthenozoospermia	Seminal plasma	1D-LC-MS/MS	CST3, AZGP1, TIMP1, SEMG1, and KLK3	Sharma et al. (2013c)
Oligoasthenozoospermia	Seminal plasma	2D-NanoLC-ESI-Q-TOF MS	NPC2, LGALS3BP, LCN1 and PIP	Giacomini et al. (2015)
Unexplained infertility	Spermatozoa	NanoLC-LTQ-Orbitrap MS	SPATA 24, ROPN1L, CRISP2, HSPA2, HSPA5, HSPB1, STIP1, CLU	McReynolds (2014)
Semen oxidative stress	Spermatozoa	1D-LC-MS/MS	HIST1H2BA, MDH2, TGM4, GPX4, GLUL, HSP90B1, HSPA5	Sharma et al. (2013b)
Semen oxidative stress	Spermatozoa	1D-LC-MS/MS	CLGN, TPPII, DNAI2, EEA1, HSPA4L, SERPINA5	Ayaz et al. (2015)
Semen oxidative stress	Seminal plasma	1D-LC-MS/MS	PIP, SEMG2, ACPP, CLU, AZGP1, KLK3, CST4, ALB, LTF, FN1, MIF and LGALS3BP	Sharma et al. (2013a)
Semen oxidative stress	Seminal plasma	1D-LC-MS/MS	MME	Agarwal et al. (2015a)
Semen oxidative stress	Seminal plasma	Shotgun proteomic analysis	FN1, MIF, G3BP, MUC5B	Intasqui et al. (2015)
Sperm DNA fragmentation	Spermatozoa	NanoUPLC-ESI-MS[E]	HIST1H2AH, LTF, ODF1 and SPACA4, ZPBP2	Intasqui et al. (2013a)
Sperm DNA fragmentation	Seminal plasma	NanoUPLC-ESI-MS[E]	ALB, NPC2, PATE4 and EDDM3A	Intasqui et al. (2013b)
Sperm DNA fragmentation	Seminal plasma	NanoUPLC-ESI-Orbitrap MS/MS	PSMA5	Intasqui et al. (2016)

(continued)

Table 3.1 (continued)

Condition	Sample	Methods used	Proteins identified	Reference
Sperm mitochondrial alterations	Seminal plasma	NanoUPLC-ESI-Orbitrap MS/MS	ANXA7	Intasqui et al. (2016)
Sperm acrosome defects	Seminal plasma	NanoUPLC-ESI-Orbitrap MS/MS	ERP44 and GSTM3	Intasqui et al. (2016)
Diabetes	Spermatozoa	2D-DIGE-MALDI-TOF MS/MS	PIP, ODF1 and SEMG1	Kriegel et al. (2009)
Diabetes	Spermatozoa	2D-DIGE-MALDI-TOF-TOF MS/MS	SEMG1, CLU, LTF and GLB1L	Paasch et al. (2011)
Failed fertilization	Spermatozoa	2D-NanoESI-Q-TOF MS/MS	ODF2	Pixton et al. (2004)
Spinal cord injury	Seminal plasma	NanoUPLC-ESI-MS[E] and 2D-ESI-Q-TOF MS/MS	ACTCM, ACTBM, ACTGM, NRAP, ACTN3, SYNE1 and SPTA2	da Silva et al. (2013)
Spinal cord injury	Seminal plasma	Dimethyl labeling-strong anion exchange LC-MS/MS	SERPINA1, SERPINA5, A2 M, ACR, KLK2, KLK3, KLK11 and PRTN3	da Silva et al. (2016)
Epididymitis	Spermatozoa		ATP5B, TUBA1A and TUBB4B	Pilatz et al. (2014)
Androgen deficiency	Seminal plasma		AZGP1, ACPP and PIP	Milardi et al. (2014)
Smoking	Seminal plasma	2D-ESI-QTOF MS/MS	ZA2G, SODE, PTGDS, ANAX3, CALM and AIAT	Fariello et al. (2012)
Smoking	Seminal plasma	NanoUPLC-ESI-Orbitrap MS/MS	LCN2, ORM1PRELP, SCGB2A1 and CEL	Antoniassi et al. (2016)

Yamakawa et al. carried out another comparative proteomic profiling between fertile and azoospermic males (both NOA and OA) for detecting candidate biomarkers (Yamakawa et al. 2007). They identified four potential markers for OA, namely stabilin-2 (STAB2), centrosomal protein of 135-kd (CP135), Ras-specific guanine nucleotide–releasing factor 1 (GNRP), and PIP. Only one marker for NOA was identified, i.e., NPC2.

In another study, five men with NOA were evaluated, and 2,048 proteins were identified (Batruch et al. 2012). Comparing these results with fertile and PV groups from a previous study (Batruch et al. 2011), they identified a total of 2500 proteins in the seminal plasma. Of these, 18 proteins were exclusive or overexpressed in NOA compared to the fertile group, such as collagen alpha-2(VI) chain (COL6A2, previously identified as overexpressed in PV), gamma-glutamyltransferase 7 (GGT7) and sorbitol dehydrogenase (SORD). The fructose metabolism was enriched in this group, indicating hypospermatogenesis or maturation arrest, according to the authors. On the other hand, 34 proteins were underexpressed or absent in NOA (of which 29 proteins were also at higher concentration in fertile men relative to PV), mostly extracellular proteins involved with reproductive functions and glycolytic pathways, such as PGK2, L-lactate dehydrogenase C chain (LDHC), zona pellucida-binding protein 2 (ZPBP2) and epididymal sperm-binding protein 1 (ELSPBP1). SORD was the only protein exclusively observed in NOA, both in comparison with control and PV, and may thus be a potential biomarker for NOA.

To follow-up these studies and to reduce the list of proteins suggested as biomarkers of a differential diagnosis of azoospermia, Drabovich et al. selected 20 proteins for a targeted MS proteomic approach, and analyzed the seminal plasma from fertile men (about to undergo a vasectomy, $n = 12$), PV ($n = 8$) and NOA ($n = 10$) (Drabovich et al. 2011b). Sixteen proteins were capable of differentiating controls from PV (for example, LDHC, sperm-associated antigen 11B—SPAG11B and mucin-15—MUC15, all not observed in PV), 3 differentiated NOA from controls (TEX101, LDHC and bile salt-activated lipase—CEL, absent in NOA) and 11 differentiated PV from NOA (such as TEX101 and SPAG11B) (Drabovich et al. 2011). To confirm these data, the same group analyzed the seminal plasma of 119 men with normal spermatogenesis or azoospermia (Drabovich et al. 2013). Their data indicated that TEX101 expression is higher in fertile men, whereas extracellular matrix protein 1 (ECM1) levels are higher in fertile men and NOA, but highly decreased in PV. EMC1 was also analyzed by enzyme-linked immunosorbent assay (ELISA) in 159 seminal plasma samples, which revealed that a cutoff value of 2.3 µg/mL differentiates between PV and NOA with high specificity and sensitivity. Values higher than this suggest NOA, while lower values indicate OA. Moreover, it can also differentiate between PV and fertile men with 100 % of specificity and sensitivity. TEX101 was also analyzed using immunohistochemistry in testicular tissues from fertile men ($n = 5$) and men with hypospermatogenesis (HS, $n = 5$), maturation arrest (MA, $n = 5$) and SCO ($n = 5$). Lower levels of TEX101 were found in spermatocytes, spermatids and sperm from HS and MA, whereas no staining was observed in SCO. In the seminal plasma, TEX101 levels were assessed by Selected Reaction Monitoring (SRM) assay and found to be undetectable only in SCO, OA, and PV samples. Fertile men presented TEX101 levels higher than 120 ng/mL, whereas HS and MA presented levels between 5 and 120 ng/mL. Recently, this group has shown that TEX101 presents a high specificity and sensitivity to identify OA, in a noninvasive manner (Korbakis et al. 2015) These data might increase the confidence in NOA and OA diagnosis using these two biomarkers, and facilitate the prediction of TESE outcome. However, Davalieva et al. have proposed the utility of prostatic acid phosphatase (PAP) in seminal plasma, as an azoospermia marker (Davalieva et al. 2012).

To study only NOA, Freour et al. evaluated the seminal plasma proteome from 40 NOA patients, which were submitted to TESE, and then classified as NOA+ (presenting sperm in the biopsy, $n = 20$) or NOA- ($n = 20$) (Freour et al. 2013). Sixty-eight proteins were differentially expressed, of which they selected CLU, PIP and galectin-3-binding protein (LGALS3BP) as potential spermatogenesis biomarkers in the seminal plasma. LGALS3BP was further confirmed by ELISA and presented higher levels in NOA+ samples, with a concentration lower than 153 ng/mL associated with a negative TESE outcome. Using this protein expression, successful sperm retrieval in TESE was correctly classified in 65 % of cases.

Similarly, differential proteins expressed in NOA males mainly fell into the types of signal transduction, cytoskeleton, and catalytic activity, suggesting that NOA may be related with the M phase of the mitotic cell cycle at the protein level, but its specific mechanism still remains unknown (Bai et al. 2010).

3.4.2 Varicocele

Varicocele is usually defined as the abnormal dilations and blood flow reversal in the pampiniform venous plexus within the spermatic cord (Agarwal et al. 2012). This is known to be a major cause of male infertility as it impairs the countercurrent heat exchange mechanism, which causes an increase in stage-specific apoptosis of germ cells at the most susceptible stages of spermatogenesis, chronic hypoxia, and excessive production of reactive oxygen species. The disease is diagnosed in about 15 % of the adult male population and is present in almost 40 % of infertile males. In fact, it is prevalent in about 35 % of men with primary infertility and rises to about 81 % in men with secondary infertility (Esteves and Agarwal 2016). Although surgical corrections of varicocele are widely used when there are alterations in semen analysis and functional tests, its efficacy has been a subject of intense debate (Cocuzza et al. 2008). Therefore, more objective criteria for the indication of varicocele repair are needed. In this context, the study of the sperm and seminal plasma proteome has generated a great deal of interest since it not only helps in understanding the mechanisms of varicocele-related infertility, but it is also very relevant towards the identification of a varicocele that will indeed lead to infertility, and, therefore, should be treated, in contrast to the "silent" varicocele.

In this regard, in order to identify potential sperm biomarkers for varicocele, comparative proteomic studies on men affected with unilateral as well as bilateral varicocele was carried out in four different experimental settings by Agarwal and co-researchers (Agarwal et al. 2015b, c, 2016b, c). In the first, a proteomic study was carried out to compare infertile men with bilateral varicocele and fertile controls. Fifty-eight differentially expressed proteins were identified, being 7 exclusively observed in sperm from men with bilateral varicocele. Of these, tektin-3 (TEKT3) and T-complex protein 11 homolog (TCP11) were selected and validated as the key sperm biomarkers of infertility in bilateral varicocele (Agarwal et al. 2016b). The same study was performed in infertile men with unilateral varicocele and demonstrated 38 unique proteins in these patients, such as cysteine-rich secretory protein 2 (CRISP2) and arginase-2 (ARG2) (Agarwal et al. 2015b). Similarly, a comparative proteomic analysis between infertile men with unilateral or bilateral varicocele and proven fertile men revealed that glutathione S-transferase Mu3 (GSTM3), sperm protein associated with the nucleus on the X chromosome B1 (SPANXB1), protein deglycase DJ-1 (PARK7), proteasome subunit alpha type-7-like protein (PSMA8), dihydrolipoyl dehydrogenase (DLD), SEMG1, and SEMG2 could be considered as potential biomarkers (Agarwal et al. 2015c). Thus, these identifications in turn may aid urologists in providing effective solutions and a strategic rationale in the selection of patients who would most benefit from varicocelectomy. Recently the same group (Agarwal et al. 2016c) reported that 87 % of the differentially expressed proteins involved in major energy metabolism and key sperm functions were underexpressed in the varicocele group (both unilateral and bilateral). Key protein functions affected in the varicocele group were spermatogenesis, sperm motility, and mitochondrial dysfunction (Agarwal et al. 2016b).

An exciting study that compared the proteomic profiles of spermatozoa from patients with varicocele and poor sperm quality before and after varicocelectomy was carried out by Hosseinifar's group (Hosseinifar et al. 2014). The study revealed that the expression of heat shock protein A5 (HSPA5), SOD1, and ATP synthase subunit delta, mitochondrial (ATP5D) increases after varicocelectomy, demonstrating the role of these proteins with sperm quality. Camargo et al. have also analyzed the seminal plasma proteome before and after varicocelectomy (Camargo et al. 2013). They concluded that nitric oxide metabolism was enriched in the pre-varicocelectomy group, whereas important functions such as response to reactive oxygen species, gluconeogenesis, nicotinamide adenine dinucleotide-binding and protein stabilization were enriched in the post-varicocelectomy group, demonstrating a shift back to homeostasis status after the surgery.

Similarly, the study of the seminal plasma and sperm proteome in adolescents with varicocele may help the early identification of varicocele-induced testicular dysfunction. In this context, studies have shown that the seminal plasma proteome differs between adolescents without varicocele, with varicocele and normal semen analysis, and with varicocele and altered semen analysis (Zylbersztejn et al. 2013; Del Giudice et al. 2016). Furthermore, in a recent study, Del Giudice et al. validated two differentially expressed proteins between these groups and demonstrated that the 45 kDa calcium-binding protein (CAB45) is underexpressed in both varicocele groups, whereas cysteine-rich secretory protein 3 (CRISP3) is significantly more expressed in the seminal plasma of adolescents with varicocele and seminal alterations (Del Giudice et al. 2016). In order to establish the efficacy of varicocelectomy as a treatment option in adolescents with varicocele, the seminal plasma of 19 adolescents was compared pre- and post-varicocelectomy, demonstrating that 19 proteins are differentially expressed after the surgery (Del Giudice et al. 2013).

3.4.3 Proteomic Profiling with Other Conditions of Altered Semen Parameters

Asthenozoospermia (AS), defined as sperm motility of less than 40 %, is a common cause of human male infertility whose etiology remains unknown in the majority of cases. Current proteomic tools now offer the opportunity to identify proteins which are differentially expressed in AS semen samples and which may be potentially involved in infertility. Sperm motility is an important prerequisite for successful fertilization and is regulated by cyclic AMP activated protein kinase-A which phosphorylates flagella proteins like axonemal dynein and initiates motility. The study conducted by Siva et al. (2010) interestingly found that although the sperm proteins falling in the functional group of "energy and metabolism" are higher in the AS patients, the proteins involved in "movement and organization" and "protein turnover, folding and stress response" were higher in the normozoospermic samples.

In two independent studies, Martinez-Heredia et al. and Hashemitabar et al. compared the whole sperm proteome and the sperm tail proteome, respectively, between normal controls and asthenozoospermic patients. Seventeen proteins (Martinez-Heredia et al. 2008) and 14 putative protein markers (Hashemitabar et al. 2015) of AS were proposed, such as cytochrome c oxidase subunit 6B (COX6B) and heat shock-related 70 kDa protein 2 (HSPA2), identified in both studies.

In order to identify the molecular basis of sperm motility, sperm from normozoospermic men were separated into moderate-motile sperm and good motile sperm, and the proteomes from both fractions were compared. Protein tyrosine phosphatase non-receptor type 14 (PTPN14) was selected as the most important impaired protein in sperm presenting alterations in motility (Chao et al. 2011). Comparing two sperm subpopulations of normozoospermic men (nonmigrated and migrated sperm after sperm selection), it has been demonstrated that nonmigrated sperm present a similar proteome as sperm from asthenozoospermic patients, with major altered proteins involved with energy metabolism, and protein folding and degradation, demonstrating that energy dysfunction is related to impaired motility (Amaral et al. 2014b). Additionally, another study has demonstrated that Prostate and testis-expressed protein 1 (PATE1) has the same alteration pattern (both in expression and localization) in asthenozoospermic patients and aged men (Liu et al. 2015).

Targeting to unravel the PTMs of sperm proteins, Parte et al. (2012) showed that 66 phosphoproteins were differentially regulated in AS. The deregulated proteins included predominantly Heat Shock Proteins (HSPs), cytoskeletal proteins, proteins associated with the fibrous sheath, and those associated with energy metabolism.

Apart from sperm, the seminal plasma from AS patients has also been explored. In one study, 45 proteins were threefold upregulated and 56 proteins were threefold downregulated in the AS group when compared with the control (Wang et al. 2009). Further, this study found that downregulation of PARK7 protein is involved in oxidative stress in semen, as was evident in AS patients.

Globozoospermia is a severe form of teratozoospermia characterized by round-headed spermatozoa with an absent acrosome, an aberrant nuclear membrane and midpiece defects. Proteomic analysis of globozoospermic sperm have shown that proteins of the perinuclear theca (PT), which has been related to acrosomal development, are significantly decreased as compared to controls (Alvarez Sedo et al. 2012). The alterations observed during early acrosome biogenesis in globozoospermia are due to anomalous development of Golgi-derived proacrosomic vesicles, failure of PT proteins to properly associate with the nuclear surface and significant deficiencies in specific PT components that are necessary for proper acrosome formation, implantation and expansion over the spermatid nucleus (Alvarez Sedo et al. 2012). Further, Liao et al. identified 9 upregulated and 26 downregulated proteins in round-headed spermatozoa compared with normal spermatozoa (Liao et al. 2009). The differentially expressed proteins were proposed to play important roles in a variety of cellular processes and structures, including spermatogenesis, cell skeleton, metabolism, and spermatozoa motility. Similarly, investigations of sperm isolated from necrozoospermic patients have demonstrated that 178 proteins are differentially expressed (Luo et al. 2008). Six proteins were found to be completely absent in the necrozoospermic spermatozoal map which may be associated with the development of necrozoospermia.

Oligoasthenoteratozoospermia (OAT) is also considered to be one of the most common causes of male factor infertility wherein many underlying factors have been hypothesized, including: chromosomal abnormalities, asymptomatic infection, mitochondrial abnormalities, environmental pollutants, subtle hormonal changes, age and functional post-testicular organ alteration (Herwig et al. 2013). In an attempt to investigate the role of seminal plasma proteins in the related male pathophysiological disorders, Sharma et al. (2013c) performed a comparative proteomic analysis of seminal plasma in men diagnosed with oligozoospermia or OAT. They have identified 1 downregulated protein (cystatin-C precursor—CST3) in the OAT group, and 2 upregulated proteins of both oligozoospermic (AZGP1 and metalloproteinase inhibitor 1 precursor—TIMP1) and OAT (KLK3 isoform 1 preprotein and SEMG1 isoform b preprotein) samples. The functional analysis of these proteins suggested that most of the identified proteins were of extracellular origin and that biological regulation is the major affected process. Another study was carried out to compare the seminal plasma proteome between normozoospermic and OAT men, and identified two underexpressed (NPC2 and LGALS3BP) and two overexpressed (lipocalin-1—LCN1, and PIP) proteins in OAT (Giacomini et al. 2015). A study was then conducted to compare the protein profile of seminal plasma from infertile men with OAT due to oxidative stress with that of healthy, fertile men to determine the proteins that are indicative of infertility (Herwig et al. 2013). The study identified a total of 2489 proteins from seminal plasma, and twenty-four proteins were determined as ≥ 1.5-fold up-regulated in the infertile idiopathic OAT (iOAT) males as compared with the fertile controls. Pathway analysis of the proteins identified solely in iOAT patients revealed an enrichment of the glycerolipid metabolism pathway. Understanding the proteins that are indicative of oxidative stress and inflammation will assist in the selection of new drugs and therapy for the treatment of iOAT infertility.

3.4.4   Unexplained Male Infertility

Unexplained male infertility is another extreme condition wherein normal semen parameters in infertile men are observed on multiple occasions with no obvious physical or endocrine abnormality. This condition is also accompanied in the absence of any female factor abnormality. In addition to erectile problems and coital factors, immune causes and dysfunctional sperm may contribute to such a condition. Unexplained male infertility may be a significant contributor to unsuccessful infertility treatment, with increasing evidence supporting an association between the quality of the male gamete and live birth.

Studies using proteomic approaches have identified important proteins associated with male fertility and dysfunction. Alterations in the sperm proteome may be

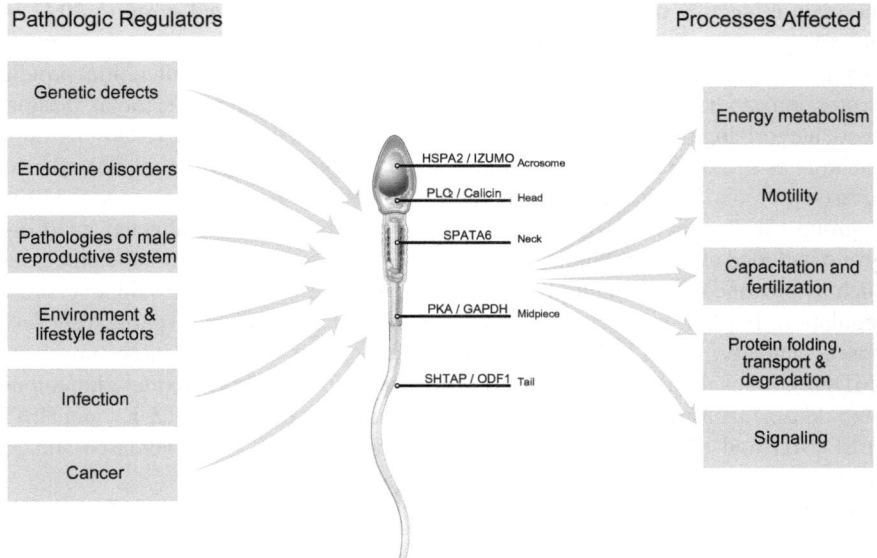

Fig. 3.1 Representative diagram of regulation of sperm proteome

a contributing factor to a decline in male fertility (Fig. 3.1). Recently, in order to understand the proteomic changes that occur in unexplained infertility, McReynolds et al. evaluated the proteome of the male gamete of normozoospermic males based on current semen analysis and investigated the relationship to in vitro fertilization (IVF) outcomes in donor oocyte cycles (McReynolds et al. 2014). The study reported 49 proteins with statistically significantly differential abundance in relation to blastocyst development (>1.5-fold). Of these, 29 proteins were underexpressed in men of the poor blastocyst development group, whereas 20 proteins were over-expressed in this group. Proteins with altered expression involved in spermatogenesis include spermatogenesis-associated protein 24 (SPATA 24), ropporin-1-like protein (ROPN1L), and CRISP2 which were lower in abundance. HSPs and heat-shock-related proteins were found to be more abundant in the sperm samples that contributed to poor blastocyst development. These proteins include HSPA2, 78- HSPA5, heat shock protein beta-1 (HSPB1), stress-induced phosphoprotein 1 (STIP1), and CLU. HSPs are involved with the prevention against oxidative stress, which is a proven factor responsible in mediating significant changes in the expression profile of several proteins, and thus may play a role in causing idiopathic male infertility (Agarwal et al. 2014).

3.4.5 Semen Oxidative Stress

Inference can be drawn by making a detailed proteomic analysis of both spermatozoal and seminal plasma proteins in association with semen oxidative stress, one of the main molecular mechanisms of male infertility and observed in up to 40 % of infertile men. In an attempt to identify protein alterations that occur as a result of increased production of reactive oxygen species (ROS), extensive differential protein expression studies were carried out, involving both sperm and seminal plasma, post-categorizing males into ROS positive (ROS+) or ROS negative (ROS−) groups (Sharma et al. 2013a, b). In one study, about 47 overexpressed and 27 underexpressed proteins were identified in sperm from the ROS+ group compared to the ROS − group (Sharma et al. 2013b). The major metabolic processes affected in ROS+ group included energy metabolism and regulation, carbohydrate metabolic processes such as gluconeogenesis and glycolysis, protein modifications and oxidative stress regulation. Further, some potential markers of oxidative stress in sperm were proposed, namely, HIST1H2BA, mitochondrial malate dehydrogenase precursor (MDH2), transglutaminase 4 (TGM4), phospholipid hydroperoxide glutathione peroxidase 4 isoform A precursor (GPX4), glutamine synthetase (GLUL), and HSPs (HSP90B1 and HSPA5). In another study, 6 protein spots were increased and 25 were decreased in the ROS+ group (Hamada et al. 2013).

Similarly, a proteomic profiling of seminal plasma (Sharma et al. 2013a) showed that upregulated proteins in the ROS+ group included PIP, SEMG2 precursor, and ACPP short isoform precursor. The proteins downregulated in the ROS− group included the CLU preprotein, AZGP1, KLK3 isoform I preprotein and SEMG2 isoform a preprotein. Moreover, some proteins unique to ROS+ group were identified, namely, the cystatin-S precursor (CST4), albumin (ALB) preprotein, LTF precursor-1 peptide and KLK3 isoform 4 preprotein. Finally, some proteins absent in ROS+ group but present in ROS− group, such as Fibronectin 1 isoform 3 preprotein (FN1), macrophage migration inhibitory factor (MIF), and LGALS3BP were also reported.

Later, the same authors divided 42 infertile men according to seminal ROS levels (low, medium, and high), and compared both their sperm and seminal plasma proteomes with that from fertile men. In seminal plasma, they demonstrated that metallo-endopeptidase (MME) protein had a more than two-fold increase in these groups. Moreover, proteins related to protein folding and degradation were differentially modulated in the seminal plasma of infertile groups, compared to fertile controls (Agarwal et al. 2015a). In sperm, almost 200 proteins were differentially expressed in the infertile groups (Ayaz et al. 2015).

In another study, targeted to determine the seminal plasma proteome in association with semen lipid peroxidation levels in men with normal semen parameters (Intasqui et al. 2015), 23 proteins were absent or underexpressed, and 71 proteins were exclusive or overexpressed in the seminal plasma of men with high lipid peroxidation levels. The main enriched sperm functions in association with seminal lipid peroxidation included unsaturated fatty acids biosynthesis, oxidants and antioxidants activity, cellular response to heat stress, and immune response. Mucin-5B (MUC5B) was suggested in this study as a potential biomarker of semen oxidative stress.

Oxidative stress may lead to male infertility especially due to sperm functional alterations. The most important observed alteration is sperm DNA fragmentation, which is associated with low pregnancy rates, both in vivo and in vitro. To identify the seminal plasma alterations related to high sperm DNA fragmentation, two separate studies were performed, comparing the seminal plasma proteome of men presenting low sperm DNA fragmentation with that of men with high sperm DNA fragmentation. In one study, 30 proteins were differentially expressed in the high sperm DNA fragmentation group, which showed an increase innate immune response and decrease lipoprotein remodeling and regulation (Intasqui et al. 2013b). Proteins related to prostaglandin biosynthesis and fatty acid binding were also overexpressed in these men (Intasqui et al. 2016). Using different bioinformatics approaches, the authors have suggested the proteasome subunit alpha type-5 protein (PSMA5) as a seminal biomarker of high sperm DNA fragmentation (Intasqui et al. 2016). The sperm proteome in patients presenting high sperm DNA fragmentation was also evaluated, and 23 proteins were exclusively or overexpressed, which were mostly related to cellular detoxification and acrosome assembly (Intasqui et al. 2013a).

In addition to sperm DNA fragmentation, oxidative stress may also impair the sperm mitochondrial activity and acrosome integrity. Aiming at analyzing the association between the seminal plasma proteome with these alterations, Intasqui et al. divided the groups into low and high sperm mitochondrial activity and low and high acrosome integrity. In this case, annexin A7 (ANXA7) was selected as biomarker of impaired sperm mitochondrial activity, whereas endoplasmic reticulum resident protein 44 (ERP44) and GSTM3 were proposed as seminal biomarkers of acrosome defects (Intasqui et al. 2016).

3.4.6 Other Infertility Conditions

Several other male infertility conditions were also evaluated by proteomic analysis. The sperm protein expression of men with diabetes or obesity was compared with fertile men, and seven proteins were associated with type 1 diabetes, whereas nine differentially expressed proteins were observed in obesity (Kriegel et al. 2009). In another study, SEMG1, CLU, and LTF were overexpressed in sperm from three different groups (type-1 diabetes, type-2 diabetes and obesity), while beta-galactosidase-1-like protein (GLB1L) was significantly underexpressed in these patients (Paasch et al. 2011).

The sperm proteome of one patient who present failed fertilization after IVF was also compared to fertile donors, demonstrating an increased expression of PIP and outer dense fiber protein 2 (ODF2) (Pixton et al. 2004). In spinal cord injured men, a differential seminal plasma proteome was reported when compared to controls, and it varied according to the applied semen retrieval method (penile vibratory stimulation or electroejaculation) (da Silva et al. 2013). In these men, an in-depth characterization of seminal plasma proteome was later performed, which enabled the identification of 2800 proteins. This analysis demonstrated an increased immune response in these patients and failure in prostate gland activity (da Silva et al. 2016). Additionally, in men with epididymitis, 35 sperm proteins were underexpressed compared with controls, such as ATP synthase subunit beta, mitochondrial (ATP5B), tubulin alpha-1A chain (TUBA1A) and tubulin beta-4B chain (TUBB4B) (Pilatz et al. 2014).

To identify altered proteins in the seminal plasma of men with androgen deficiency, Milardi et al. evaluated 10 men with proven fertility and 20 men with secondary post-neurosurgical hypogonadism, to study only men presenting drastic reduction in circulating testosterone levels. Ten men were also analyzed after one week of a 6-month treatment with testosterone enantate (250 mg, every 3 weeks). From the 61 identified proteins in the fertile men, 33 were absent in all hypogonadic men, of which 14 were also identified in post-therapy patients, such as AZGP1, ACPP, and PIP. Absent proteins were mostly involved with catalytic and binding activities, and hydrolase was the most impaired enzyme by T deficiency (Milardi et al. 2014).

In the seminal plasma of smokers, it has been demonstrated 33 differentially expressed proteins compared to nonsmokers, which were mostly involved with antigen processing and presentation and regulation of immune response (such as, prostaglandin secretion, and acute inflammation). Interestingly, these alterations were also associated with decreased sperm function, in particularly, with high sperm DNA fragmentation, low sperm mitochondrial activity, and acrosome damage (Antoniassi et al. 2016). In smokers with varicocele, 20 differentially expressed proteins were observed in the seminal plasma, as well as increased sperm DNA fragmentation and decreased sperm mitochondrial activity (Fariello et al. 2012).

Research in the field of sperm and seminal plasma proteome is on the rise and the search for suitable and potential biomarkers of male infertility is still ongoing. Although remarkable developments have been achieved in the field of andrology in the recent years which have significantly improved our understanding of sperm physiology, there are still many questions that remain unanswered. Intriguingly, different sets of proteins have been proposed as biomarkers in different conditions of reduced fertility and/or infertility and there is minimal overlap of some of the identified proteins that are found to be differentially expressed in male infertility by independent studies. One of the plausible reasons for this heterogeneity between independent studies is that different proteomics techniques and their combinations were used. It is therefore speculated that the recent development of high-throughput MS-based technology will allow more detailed investigation of the proteomes of interest, among which will result in more reproducible results in the future.

Chapter 4
Role of Proteomics in Female Infertility

Nirlipta Swain, Caroline Cirenza and Luna Samanta

Deterioration of the female fertility potential has been one of the major areas of research since decades. However, the mysteries and uncertainties still remain largely elusive. Since it contributes an equal 50 % for a successful reproductive outcome, assessment of the exact factors of female infertility might help for the accomplishment of a live birth. Based on previous reports, it is seen that the causes and conditions of female infertility can be both congenital as well as acquired (Abrao et al. 2013). Nevertheless, several such factors unraveled till date are still doubted to be a causative or a confounding factor. The above obfuscates the intricacies involved in both regulation and maintenance of fertility, as well as in dysfunctional conditions leading to infertility.

In couples facing the complications of infertility (both primary and secondary), diagnosis reveals that in 20–30 % of cases, both male and female partners are defective, while in approximately 30 %, either one of the partners is found to be responsible (Agarwal et al. 2005). Moreover, 10–15 % cases remain idiopathic with the standard fertility parameters being normal. Currently, 48.5 million females within the ages of 20–44 years have been examined and found to have some form of infertility (Mascarenhas et al. 2012). All the known anomalies leading to female

N. Swain · L. Samanta (✉)
Department of Zoology, Ravenshaw University, Cuttack, India
e-mail: lsamanta@ravenshawuniversity.ac.in

N. Swain
e-mail: nirliptaswain@gmail.com

C. Cirenza
The Medical School University of São Paulo, São Paulo, Brazil
e-mail: carolcirenza@gmail.com

© The Author(s) 2016
A. Agarwal et al., *Proteomics in Human Reproduction*,
SpringerBriefs in Reproductive Biology, DOI 10.1007/978-3-319-48418-1_4

infertility can be categorized into genital, endocrinological, developmental, and general factors. Recently, it has been reported that out of different causative infertility factors in female partners, about 25 % are ovulatory disorders, 20 % cases are due to tubal damage from pelvic infections or endometriosis, and the remaining 10 % are uterine or peritoneal disorders, such as fibroids and endometriosis (Amoako and Balen 2015).

Clinical evidences have linked numerous proteins to be involved at specific or all stages of oogenesis and embryo development, starting from proper oocyte formation to embryo implantation and maintenance. Thus, protein profiling using a proteomic approach would help to not only detect the malfunctioned proteins in infertile females, but also decipher the crosstalk among these proteins in the physiological scenario. In male partners, proteomics studies have been undertaken at a cellular level (spermatozoa), tissue level (testicular tissue), as well as of different fluids (epididymal fluid and seminal plasma). Similarly, for females also protein profiling has been done at a cellular level (on rodent oocytes), tissues, (ovarian and endometrial tissue) and of different fluids (follicular fluid, uterine fluid, and cervico–vaginal fluid). Moreover, proteomics studies have identified differential expression of proteins in female factor disorders, such as endometriosis, polycystic ovary syndrome (PCOS), and pre-eclampsia.

The present chapter focuses on highlighting the literature that have identified fertility-related proteins in the reproductive microenvironments of females, such as endometrial fluid, peritoneal fluid, and follicular fluid, as well as reproductive tissues and serum. These proteins can be taken as potential biomarkers that could be used for diagnosis for female factor disorders. Furthermore, these proteins can be targeted for therapeutic treatments to rectify the underlying dysfunctional molecular mechanisms in infertility-related disease conditions.

4.1 An Overview on the Physiological Scenario of Female Reproductive Function

The follicles in the ovaries (female gonads) are considered as functional units, being comprised of oocytes and surrounding somatic granulosa and cumulus cells (Seli et al. 2014). The reproductive function in human females is cyclic with a periodic expulsion of mature oocyte for fertilization (ovulation). The oocyte regulates follicular development and maturation (folliculogenesis), and neuroendocrine changes within the hypothalamic–pituitary–ovarian axis, causing a monthly rupture of an ovarian follicle and extrusion of an ovum into the fallopian tube (Stricker et al. 2006). Following ovulation, reorganization of the collapsed follicle takes place to form the corpus luteum, which comprises primarily luteinized granulosa cells and thecal cells. In a synchronous action, the endometrial lining of the uterus thickens with extensive structural modifications to provide the necessary support for implantation (Mihm et al. 2011). If the mature oocyte fails to fertilize, the endometrial lining is released out of the body as menstrual discharge (Yang et al. 2012). The entire ovarian cycle involves, thus, the chronological occurrence of follicular and luteal phases, being regulated by multiple endocrine signals, as is marked in the ample fluctuations in gonadotropin, sex hormones, and other metabolites in serum concentrations (Rojas et al. 2015).

Post-insemination, there is a rapid transport of spermatozoa through the vagina, cervix, and uterus, finally reaching the fallopian tube. It has been reported that in humans, semen is deposited in the anterior vagina near the cervical os during coitus, where it coagulates to form a seminal gel (Suarez and Pacey 2006). Within a few minutes following vaginal deposition, the sperm migrates from the seminal pool into the cervical canal (Sobrero and Macleod 1962). The encountered cervical mucus not only navigates the path of sperm transit through the cervix (Chretien 2003), but also acts as a selection media that screens only the normal, vigorously motile sperm to penetrate further (Katz et al. 1990).

The human uterine cavity, being relatively small of only a few centimeters in length, is covered within 10 min by sperm with a swimming speed of 5 mm/min (Mortimer and Swan 1995). As the sperm traverses the uterus, the movement is likely aided by pro-ovarian contractions of the myometrium. After crossing the utero-tubal junction, the sperm settle to form a tubal population in the tubal isthmus. In contrast to other animals, human females lack a morphologically distinct sperm reservoir in the fallopian tube. Nevertheless, a functional reservoir is established in the tubal isthmus, which ensures prolonged availability of sperm in the fallopian tube with only a few sperm at a given time being capable to reach the ampulla (Suarez and Pacey 2006).

In fact, reports propose that motile sperm reach the fallopian tube within an hour (Kupker et al. 1998) and is capable of fertilizing an ovulated ovum even after 5 days of insemination (Wilcox et al. 1995). As the sperm advances further beyond the tubal reservoir, it completes its maturation process and undergoes fusion with the ovulated ovum in the fallopian tube. Post-fertilization, the zygote develops into the blastocyst stage in the fallopian tube and gradually descends to the uterus, where it is implanted and undergoes further fetal development (Carson et al. 2000).

Therefore, the posterior part the vagina is the site of semen deposition, and the cervical canal acts as a traversing tract for sperm transit. In the anterior part of the female reproductive tract, the ovaries are the sites of oocyte maturation, and the fallopian tubes become the sites of fertilization. The uterus present in the midway carries the developing embryo till birth. Moreover, apart from serving as a guiding tract for mediating sperm transport and fertilization, the female reproductive system in humans also sustain molecular interaction for regulating sperm maturation (Suarez 2016) and embryonic implantation (Carson et al. 2000). It has been reported that ejaculated sperm undergo physiological priming during their arduous journey through the female reproductive tract. Capacitation is a continuous process that begins from insemination until the sperm reaches the vicinity of the oocyte and is marked as a preparatory phase for mediating acrosomal reaction (Fraser 1998).

Hyperactivation is a coordinated change in the flagellar beating that occurs in the tubular reservoir and triggers the release of sperm from the latter (Suarez and Pacey 2006). It is further required to enable the sperm to proceed toward the oocyte and penetrate through the vestments. Since all these events have to be not only proper but also timely, the activities of the entire female reproductive system is intricately regulated by a cohort of signaling proteins. The latter is configured to effectively carry out the maturation of both the gametes, along with mediating fertilization, and subsequent embryonic development. In this regard, proteomic profiling of all the constitutive proteins is imperative.

4.2 Mature Murine Oocyte as a Reference for Understanding Human Oocyte Maturation

The ovulated mammalian egg is transcriptionally dormant being arrested at the metaphase II (MII) stage of germinal meiosis (Yurttas et al. 2010). In fact, it has been reported that post-birth, oocytes are present at the diplotene stage in the prophase of meiosis I and continues further development just a day prior to ovulation (Cao et al. 2012). Hence, during mammalian oogenesis, oocyte maturation gets arrested sequentially at the dictyate (meiotic prophase) or germinal vesicle (GV) stage and the metaphase II (MII) stage. The ovulated, mature MII oocytes can be proposed to store a cohort of maternal proteins that were synthesized during oocyte growth, and can be implicated to regulate both fertilization and embryonic development (especially early embryogenesis). Because no such protein characterization studies have been carried on human oocytes, inference needs to be drawn from similar kinds of proteomic studies performed with mouse oocytes.

The comparative proteomic analysis by Wang et al. revealed that the GV oocyte predominantly contains the metabolism-related proteins needed for supporting oocyte maturation. Whereas, the overexpression of DNA damage and repair-related proteins was observed in the proteome of MII oocytes, with the exclusive accumulation of proteins involved in epigenetic programming and chromatin remodeling in the latter (Wang et al. 2010). Wang's group found 2781 proteins in the GV oocyte stage, 2973 proteins in the MII oocyte stage, and 2082 proteins in the fertilized oocyte (zygote), which suggests a differential expression of proteins specific to any particular developmental stage. Furthermore, an increased translational activity is observed in MII oocytes, with increased abundance of many proteins related to meiosis, fertilization, and early embryonic development (Cao et al. 2012). Moreover, the proteins have been detected to be post-translationally modified, which controls their activity. Zona pellucida (ZP) proteins, the most abundant protein in the oocyte, are known to be the main component involved in egg–sperm interaction, and are seen to be glycosylated (Zhang et al. 2009). Nucleoplasmin 2 (NPM2), which is a proven maternal-effect protein that plays an important role in early embryonic development, is seen to be phosphorylated in MII oocytes. In contrast, adenylosuccinate synthase (ADSS), an enzyme of the purine metabolism pathway is found to be dephosphorylated during oocyte maturation (Cao et al. 2012).

Studies on the identification of such putative maturation-related proteins may help us to investigate the "human oocyte factor" responsible for causing the oocyte maturation arrest observed at various stages of the cell cycle following both in vivo and in vitro ovarian stimulation (Mrazek and Fulka Jr 2003; Beall et al. 2010; Levran et al. 2002). Furthermore, the proteomics analysis of denuded MII oocytes by Pfeiffer et al. characterized only the oocyte-specific proteins barring any cumulus proteins. This work has particular relevance to assisted reproductive technologies (ART), where denuded mature oocytes are taken post-oocyte preparation (Pfeiffer et al. 2011). However, a bidirectional communication exists between the oocytes and the surrounding cumulus granulosa cells, which is essential for the development of oocytes and ovarian follicles. In a study on murine cumulus-oocyte complex (COC), a total of 156 proteins were identified to play a role in the ovarian follicular development (Meng et al. 2007).

The above data can serve as a reliable reference for the proteomic studies on COCs in pathological conditions, such as PCOS, chronic anovulation and luteinized unruptured follicle (LUF). A similar kind of protein profiling was done by Hamamah et al. who worked with human cumulus cells (CC) and found dissimilarities in the protein pattern in CC, based on the type of ovarian stimulation protocols the oocyte was subjected to (Hamamah et al. 2006). Furthermore, coupling of surface labeling with oolemal proteomics can highlight the proteins involved in the sperm–egg interaction (Calvert et al. 2003; Yurttas et al. 2010). Additionally, the proteomics analysis of oocyte-restricted proteins required for early embryogenesis can be targeted to discover novel oocyte-reprogramming proteins (Yurttas et al. 2010; Pfeiffer et al. 2011). This would help us to understand and explain the underlying mechanisms for transition from maternal to embryonic control during early development.

4.3 Follicular Fluid as a Biomarker of Oocyte Health and Quality

The developing oocyte grows and matures in an ovarian follicle, and later provides the required biological niche inside the ovary. During the antral stages of folliculogenesis, a fluid-filled cavity termed the antrum is developed inside the follicle, which gets filled with follicular fluid (FF). FF is a selected ultrafiltrate of plasma that has diffused from thecal capillaries. FF is further modified by secretion and uptake of specific components by the granulosa and thecal cells within the follicle itself. Since it provides the biochemical milieu surrounding the oocyte, FF ensures the proper development of the oocyte and determines the quality of the subsequent fertilization and embryo (Benkhalifa et al. 2015). FF acts as medium for intrafollicular signaling, by which signaling mediators are transported between the oocyte and follicular cells within the follicle, which is required for the acquisition of developmental as well as fertilization competence by the oocyte. FF is composed of several other factors apart from proteins and peptides, such as hormones, growth factors of the Transforming Growth Factor-beta (TGF-beta) superfamily and others, interleukins, anti-apoptotic factors, antioxidants, sugars, and prostanoids (Revelli et al. 2009). Being easily available as it gets aspirated together with the oocyte at the time of ovum retrieval performed during ART, FF can serve as a potential biomarker of oocyte health and subsequently, as an assessment of the success of ART.

In 1996, Spitzer et al. used 2D-based approach for the first time to study protein composition of follicular fluid (Spitzer et al. 1996). A proteomics study on human FF (HFF) from follicles at different developmental stages suggested a differential and stage-specific expression of HFF proteins involved in oocyte maturity (Liu et al. 2007). Angelucci et al. found that the HFF is mainly comprised of acute phase proteins (APP) and antioxidant proteins, which suggests that ovulation may be considered as an inflammatory event (Angelucci et al. 2006). Further, Hanrieder et al. carried out an in depth protein analysis of HFF and identified cell-adhesive proteins, such as vitronectin and Mac 2 binding protein, as well as proteins involved in hormone secretion regulation during the foliculogenetic process (Hanrieder et al. 2008). A sub-proteomic analysis of HFF identified about 246 proteins, the majority of which belonged to coagulation and immune response pathways (Twigt et al. 2012).

The first extensive characterization of HFF proteins was done by Ambekar et al. where they identified 480 proteins, out of which 320 were reported for the first time (Ambekar et al. 2013). Subcellular localization data revealed that the majority of proteins were extracellular, while the presence of intracellular proteins can be attributed to be present due to cellular apoptosis occurring while follicular development. HFF proteins are shown to be mainly involved in enzyme catalysis, transporter activity, complement activity, cell adhesion, receptor signaling, and the constituents of extracellular matrix (ECM). Similarly Bianchi et al. performed a functional characterization of 43 unique proteins (Bianchi et al. 2013). The pathway analysis shows that these proteins were mainly involved in the inflammation, complement and coagulation cascades, response to wounding, protein–lipid complex/lipid metabolism and transport, and cytoskeleton organization. In another study, it was seen that a total of 503 proteins were found in HFF, which was dominated by the albumin and immunoglobulin family proteins (Bayasula et al. 2013). Furthermore, this study demonstrated a differential proteome between FF of fertilized oocytes, compared to FF that contained oocytes that failed to be fertilized. For instance, heparin sulfate proteoglycan perlecan protein was overexpressed in the fertilized group.

During folliculogenesis, the permeability of follicles to HFF proteins in general increase, which is marked by notable similarities between protein composition of HFF and blood plasma/serum. However, although FF is generally a filtered product of plasma, yet studies have revealed a differential protein profile of FF as compared to plasma (Jarkovska et al. 2010). Quantitative and qualitative compositional changes in HFF in comparison to serum suggest that blood plasma constituents are selectively transferred through the blood follicular barrier via theca capillaries (Schweigert et al. 2006). Some proteins identified to be selectively transported are haptoglobins and transthyretin (TTR). A total of 12 proteins were seen to be differentially present in HFF in comparison to serum (Jarkovska et al. 2010). The protein groups showing significant differential expression in HFF as compared to serum are involved in the complement system-mediated immune defense, acute phase response, transport, blood coagulation, and lipid metabolism.

Moreover, several paired comparison studies were also undertaken to examine protein alterations in FF at different conditions. A marked absence or weak expression of several proteins in immature follicles was observed when compared to the protein profile of paired mature follicles collected from the same patients (Spitzer et al. 1996). This may reflect the physiological variation, and serve as a biomedical marker of follicular maturity. In a concomitant study, using an advanced proteomic approach, several proteins were identified to be potential biomarkers of good versus poor responders in matched pairs of IVF patients (Estes et al. 2009). In order to validate specific proteins that might emerge as potential candidates for diagnostic markers of oocyte quality, a comparative proteome profile of HFF was evaluated between healthy younger (20–32 years old) and older (38–42 years old) women (Hashemitabar et al. 2014). Twenty-three protein spots showed a significant variation between the two groups. Moreover, downregulation of five proteins, namely serotransferrin, hemopexin precursor, complement C3 and C4, and kininogen in older women HFF could be linked to reproductive aging. All such studies corroborate the importance of HFF proteins in regulating the follicular microenvironment, and their biological role as biomarkers of follicle and/or oocyte quality, and subsequent embryo vitality.

4.4 Evaluation of Uterine Status for Successful Implantation

Even after a successful fertilization, failure in the implantation of the embryo, would also result in infertility, and is a rate-limiting step of ART success. The endometrium is the mucosal lining of the uterus, and acts as a bed for receiving the fertilized ovum. There must be a synchrony between the dividing blastocyst and the well-nourished fully matured endometrium for proper implantation to occur. Endometrium undergoes a cyclic remodeling with every menstrual cycle, which is characterized by erosion of the upper *stratum functionalis* layer and subsequent replacement with regenerated tissue from the underlying *stratum basalis* layer. Such a monthly process of periodic growth, differentiation, desquamation, and regeneration of endometrial layers is regulated by the ovarian steroidal hormones estrogen and progesterone as well as other hormones, cytokines and chemokines. The menstrual cycle is divided into four distinct phases of marked physiological change, namely the menstrual phase, proliferative phase, ovulatory phase, and secretory phase. Shortly after ovulation, in the secretory phase, the endometrium acquires a functional and transient progesterone-dependent status, allowing blastocyst adhesion. This spatially and temporally restricted period of the mid-secretory phase varies from 12 h to 2 days and is known as the "window of implantation" (WOI) (Paria et al. 2000). In order to understand the molecular mechanism involved in this transition from nonreceptive to receptive status of uterus, a comprehensive proteomics study of endometrial tissue and uterine fluid is highly essential.

A comparative proteomic profiling of the proliferative and secretory phases of the human endometrium revealed information about the stage-specific production of proteins. The proliferative phase is the period of extensive cellular division as the tissue gets reconstructed following the menstruation phase, and the secretory phase is the time of reorganization and differentiation of endometrial tissue. An earlier study found the presence of two anti-apoptotic proteins, glutamate NMDA receptor subunit zeta 1 and FRAT1 exclusively in the secretory proteome, which suggests their utility as a biomarker (DeSouza et al. 2005). Byrjalsen et al. found that the most expressed proteins in the proliferative phase are mainly cytoskeletal proteins (vimentins, keratin, tropomyosin and tubulin), proliferating cell nuclear antigen, and β-galactoside binding lectin (Byrjalsen et al. 1995). Meanwhile in the secretory phase of the endometrial tissue, proteins involved in energy metabolism such as creatine kinase chain B and an isocitrate dehydrogenase-homologous protein were identified. An extensive characterization of the differentially expressed proteins was done by Rai's group (Rai et al. 2010a). They also proposed that more proteins are downregulated than upregulated in the secretory phase, in comparison to the proliferative phase. The downregulated proteins include metabolic proteins, such as NADH dehydrogenase (ubiquinone) Fe–S protein 1 and mitochondrial aldehyde dehydrogenase; signaling proteins involved in cell cycle control, cell proliferation and anti-apoptosis; proteins involved in RNA biogenesis, such as heterogeneous nuclear ribonucleoprotein C, and B23 nucleophosmin; and the abundant endometrial Heat-Shock Proteins (HSPs).

However, the structural proteins (vinculin and F-actin capping protein b-subunit) involved in cell adhesion, maintenance of cell morphology, cell proliferation and cell migration are upregulated, which suggests the association of these processes with proliferative-to-secretory phase transition in endometrium. Further, it was seen that during implantation, an anticoagulatory environment can be suggested as evidenced by upregulation of annexin V and downregulation of fibrinogen γ during the secretory phase.

The mid-secretory phase is actually the phase when the endometrium is receptive to the developing embryo, so several studies have targeted the endometrial tissue at this stage for a comparative proteomic profiling with that of the nonreceptive stage. Proteins seem to be upregulated in the mid-secretory phase endometria includes annexin V, peroxidoxin 6, alpha1-antitrypsin (AAT), and creatine kinase, in contrast to transferrin, calreticulin, adenylate kinase isoenzyme 5, and the beta chain of fibrinogen in the proliferative phase endometria (Parmar et al. 2009). Another study associated the different identified upregulated proteins to two biological pathways, namely JNK and EGF signaling pathways—while the former is related to cell organization and assembly, the latter is linked to cellular differentiation, inhibition of proliferation, apoptosis, migration, and adhesion (Chan et al. 2009). Phenotypic changes of the receptive endometrial tissue during mid-secretory phase, being termed as "plasma membrane transformations" are characterized by not only differential production of proteins, but also presence of different isoforms of the same protein. These recommend the occurrence of post-translational modifications of proteins during the transition from proliferative-to-secretory phase. In fact, a recent study proposes that N-glycosylated ENPP3 may be a candidate biomarker for the normal mid-secretory phase of the endometrium (Hood et al. 2015). Instead of taking whole-endometrial-tissue homogenates, Hood's group undertook histological assessment of the glandular epithelium and surrounding stroma. They concluded that out of 1224 proteins of glandular epithelia, 318 proteins were differentially expressed between proliferative and secretory phases, while for 1005 proteins of stromal tissue, only 19 were differentially expressed.

Immunohistochemical analysis of differentially expressed proteome patterns between pre-receptive and receptive endometria demonstrated an altered localization of annexin A4 along with its upregulation at the mid-secretory phase (Li et al. 2011b). Furthermore, annexin A2 and stathmin 1 were confirmed to be possible targets for the assessment of human endometrial receptivity and interception (Dominguez et al. 2009). Similarly, progesterone receptor membrane component 1 (PGRMC1) and annexin A6 can also be taken as important cytoskeletal markers for the acquisition of endometrial receptivity (Garrido-Gomez et al. 2014). Some of the pathways greatly regulated between receptive and proliferative phase in the endometrial tissues included those associated with "carbohydrate biosynthetic processes" and "nuclear mRNA splicing via spliceosome" (Garrido-Gomez et al. 2014) as well as "energy metabolism mediated by CKB and cell-cell adhesion" of the endometrial tissue (Chen et al. 2015). Moreover, KEGG pathway mapping revealed that downregulation of focal adhesion proteins would enhance endometrial receptivity (Chen et al. 2015).

The endometrial tissue provides a distinct repertoire of uniquely expressed proteins of the receptive uterus. However, due to its cellular complexity and variability both among individuals and among cycle stages, endometrial tissue may not be the ideal mode system for detecting biomarker proteins. In contrast, uterine fluid which can be regarded as the endometrial secretome represents the actual microenvironment for implantation. Being relatively less invasive than tissue biopsy, the presence of low levels of cellular proteins makes it much less complex than the endometrial tissue. Parmar and colleagues found the presence of transferrin, heat-shock protein 27, and AAT both in uterine fluid and endometrial tissues, implying that one of the sources of uterine fluid proteins is the endometrium (Parmar et al. 2009). A comprehensive proteomic characterization of human endometrial aspirate was done by Casado-Vela et al. (2009), while that of endometrial lavage was done by Scotchie et al. (2009). Proteomic studies demonstrate the presence of immunoglobulins, transferrin, alpha-1 antitrypsin precursor, anti-chymotrypsin precursor, apolipoprotein, haptoglobin, and hemopexin in human uterine fluid (Parmar et al. 2008). Biological functional annotations of identified uterine fluid proteins revealed that the highest percentage corresponded to proteins involved in metabolism (68.5 %), whereas 4.3 % of them were related to reproduction (Casado-Vela et al. 2009). A comparative proteomic analysis of uterine fluid collected during the mid-secretory phase and the proliferative phase shows that the most differentially expressed proteins were associated with host defense, coagulation, apoptosis regulation, and stress response (Scotchie et al. 2009). Moreover, proteins playing a role in embryonic implantation and development, namely, apolipoprotein A4 (APO A4), apolipoprotein A1 (APO A1) fragment and alpha-1 antitrypsin precursor were two to threefold more abundant in mid-secretory uterine fluid (Parmar et al. 2008). It was further derived that uterine gland has a reduced secretory activity at mid-secretory phase, since no significant increment in protein abundance is noticed in endometrial lavage collected during mid-secretory phase as compared to mid-proliferative phase (Hannan et al. 2010).

To unravel the mysteries of dysfunctional uterine activity in infertility conditions, Hannan and coworkers carried out a comparative proteomic profiling of endometrial lavages of mid-secretory and mid-proliferative phases from both fertile and infertile women (Hannan et al. 2010). It was seen that upregulated proteins of mid-secretory phase endometrial lavage, namely APO A4, activin receptor type-2B, interalpha-trypsin inhibitor heavy chain H4 (ITIH4), APO A1, and alpha-2-macroglobulin (A2M) showed differential expression in lavages from infertile patients. While the former three proteins were elevated, the latter two proteins were seen to be reduced in the endometrial lavages of infertile patients. All such studies create a foundation to further study causes of abnormal endometrial receptivity, implantation defects, and infertility. Interestingly, uterine fluid contains proteins that not only mediate the maintenance and nurturing of the preimplantation embryo, but also of the ascending spermatozoa.

Complementary to the aforesaid techniques to diagnose endometrial receptivity and uterine pathologies, an easily obtainable and completely noninvasive collectable body fluid for such kind of analysis could be menstrual blood. This contains the products obtained from expulsion of the endometrial lining of the uterus along with blood. Yang et al. carried out one such study comparing the menstrual blood proteome with that of circulating blood and vaginal fluid (Yang et al. 2012). They found 385 proteins unique to menstrual blood, which were mainly histones, ribosomal proteins, cytoskeletal elements, cytokines, and immunoglobulins. Since failure in implantation leads to menstruation, abnormal composition of menstrual blood would be a reflection of anomalies in the biochemistry or cellular composition of the endometrium. Abnormal menstruation can be a sign of underlying uterine pathology, and here again a comparison against the normal menstrual blood would serve as an effective diagnostic tool.

4.5 Proteomic Analysis of the Cervical–Vaginal Fluid as a Standard Diagnostic Tool

Cervico–vaginal fluid (CVF) is a mixture of secretions from vaginal cells, cervical vestibular glands, plasma transudate, and endometrial and oviductal fluids, lubricating the lower genital tract in females. CVF can serve as a biomarker for several reproductive functions predicting the success of a proper insemination and sperm transport for in vivo fertilization, to a timely onset of a proper labor for child birth. Cervical mucus is involved in the regulation of sperm transport, protecting the sperm against the hostile acidic vaginal secretions and conditioning it for capacitation, as well as filtering the morphologically abnormal and immotile sperm from the ejaculate. During the menstrual cycle, the CVF undergoes significant biochemical and physiological changes, which suggests of its use to study different phases of the menstrual cycle, especially to predict the timing of ovulation. Finally, CVF is an easily accessible body fluid in close proximity to tissues involved in dynamic changes associated with labor onset. The protein profiling of human CVF could be used as a diagnostic tool to understand the intricacies of human pregnancy and labor. This could be extrapolated to unravel the cause for different child birth-associated pathologies like pre-term labor and birth. Furthermore, since CVF forms the first line of defense, its study would provide an insight of different infectious diseases of the female reproductive tract.

The composition of maternal CVF changes with gestational age and vaginal health. A proteomic study on CVF yields that the major proteins in CVF can be grouped as blood transport proteins, structural proteins, proteins involved in fatty acid metabolism, calcium-binding proteins, anti-inflammatory cytokines, proteinase inhibitors, and enzymes involved in oxidative stress defense pathways (Di Quinzio et al. 2007). Localization distribution shows that most of the CVF proteins are either secreted/extracellular (43.3 %) or cytosolic (43.3 %) in nature (Tang et al. 2007). An extensive characterization of CVF proteome by Dasari's group represented that about 32 % were metabolic proteins, 22 % were immune response-related proteins, while the rest of the proteins were associated with cell differentiation (11 %), transport (9 %), cell organization (8 %), enzyme regulation (6 %), signal transduction (3 %), and cell proliferation (3 %) (Dasari et al. 2007). Concomitantly, Shaw and colleagues did a distinct categorization of CVF proteins into five subgroups, namely, host defense, proteolysis, enzyme inhibition, cellular adhesion, and the cytoskeleton (Shaw et al. 2007). Zegel et al. classified the CVF proteins into the following categories: protein metabolism and modification (19 %), immunity and defense (13 %), developmental process (9 %) and signal transduction (9 %) (Zegels et al. 2009). A comparative protein profiling of the CVF, serum, and amniotic fluid demonstrated that 77 proteins are unique to CVF, while 56 and 17 CVF proteins also occur in serum and amniotic fluid, respectively. Since most of the CVF proteins were seen to be having a plasma origin, it could be suggestive that CVF could largely be plasma transudate (Dasari et al. 2007).

A marked change is seen in physicochemical properties of the cervical mucus (CM) that occurs cyclically in response to hormonal changes. CM secreted by the cervical glands, fills the opening of the cervix. At the beginning of the menstrual cycle, the CM is scant, thick, and viscous, while during ovulatory phase there is an increase in volume with a reduction in viscosity, under the influence of estrogen levels. Such altered mucus properties facilitate sperm penetration into the uterus during ovulation. Post ovulation, when the corpus luteum of the ovary begins to synthesize progesterone, the CM becomes thicker and stickier. This would build up an inhibitory microenvironment for sperm propagation during the nonovulatory phases of the cycle. A proteomic approach to identify phase-specific expression of CM proteins found that, although protein expression was significantly upregulated in ovulatory CM, 42, 38, and 17 exclusive proteins were, respectively identified in the pre-ovulatory, ovulatory and post-ovulatory CM (Grande et al. 2015). Some of the unique marker proteins of pre-ovulatory CM include cystatin C, clusterin, glycodelin, and metalloproteinase inhibitor 1, which are essential for coagulation of abnormal sperm and inhibition of sperm activity. Similarly for the post-ovulatory phase, specific marker proteins include triosephosphate isomerase (glycolytic enzyme), and keratins I and II, spectrin and dynein chains, which are essential for the maintenance and organization of the intermediate filament network. About 38 proteins constitutive of the CM in all the phases were mainly defense/immunity proteins. However, proteins of CM mainly represented functional categories of metabolism, immune response, and cellular transport (Panicker et al. 2010).

Temporal changes in the cervical mucus plug (CMP) in pregnant women would reflect the cellular and molecular events occurring in the uterine cervical regions for maintenance of pregnancy and parturition. CMP is the ultimate sealant of the uterine cavity during pregnancy and is physiologically and biochemically different from the cervical secretions of non-pregnant women. In comparison to the cervical mucus of non-pregnant women or CVF, some of the unique proteins of CMP include CD81 antigen and pregnancy zone protein (Lee et al. 2011). With an enrichment of proteins involved in the immune response, such as complements and neutrophil defensin, CMP is generally regarded as a critical "gate-keeper." While the core proteins of CMP, mucus glycoproteins (mucins), provide the structural framework, other proteins such as lactoferrin and lysozyme are involved in host defense. By sharing a lot of common proteins with amniotic fluid, CMP can be used for diagnosis of certain genetic and metabolic diseases of the fetus. The onset of human labor is marked by cervical ripening, fetal membrane rupture, and myometrial activation. Heng's group has validated the utility of interleukin-1 receptor antagonist (an anti-inflammatory cytokine) as a predictor of term labor (Heng et al. 2008). Moreover, the same group have identified temporal changes in MNEI, SCCA1, AnxA3, collagen type IV and albumin expression in association with term labor (Heng et al. 2010). A comparative proteomic analysis on serial CVF samples obtained from women during late pregnancy and spontaneous labor showed a differential expression of proteins involved in protease inhibition, anti-inflammatory cytokine activity, and oxidative stress defense (Di Quinzio et al. 2008). Additionally, a significant reduction of thioredoxin expression (Di Quinzio et al. 2008) and increased expression of albumin (Heng et al. 2010) was reported in spontaneous labor patients.

4.6 Proteomics in Various Infertility Conditions and Their Possible Biomarkers

Genomic studies of female gynecological tissues have been applied to elucidate potential biomarkers of various female disorders. However, this approach does not take into consideration post-transcriptional and post-translational changes of the expressed proteins. This then strengthens the utility of proteomics as a more reliable method for biomarker discovery. Identification of the unique anomalous proteins or the array of differentially expressed proteins in diseased condition would help to make a more efficient diagnosis. Furthermore, a regular evaluation of the targeted protein expression would help to do the prognosis of the recovery or progression of the disease post-treatment. A comparative protein profiling conducted on endometrial tissue and secretions, ovaries, uterine fluid, follicular fluid, and CVF collected from fertile controls and patients would lead to a better understanding of the underlying causes and the mechanistic pathways of female infertility. Out of the several female factor infertility cases, the most common causes studied by a proteomic approach include: endometriosis, PCOS, reproductive tissue cancers, as well as pregnancy-related disorders like pre-eclampsia and pre-term labor (Table 4.1).

4.7 Endometriosis

Endometriosis is a complex gynecological disorder in which the endometrium grows outside of the uterine cavity. Being characterized by severe pelvic problems and fertility-related problems, it is mostly an estrogen-dependent disease. During the menstrual cycle, the eroded normal endometrial tissue may enter the peritoneal cavity by retrograde menstruation, get implanted, and cause lesions ectopically. However, the pathophysiological mechanisms underlying this disease are still enigmatic. The "normal" endometrial cells apparently turn "endometriotic" because of inherent molecular abnormalities present in them, which can be unraveled by comparative proteomic analysis. Comparing the expression and regulation profiles of proteins found in endometriosis with normal eutopic tissues (endometrium and peritoneum), as well as with those found in the different forms of endometriosis (i.e., peritoneal endometriosis, endometrioma, and adenomyoma), would demonstrate proteins that could be responsible for the onset and progression of endometriotic implants. Additionally, these specific implant proteins could be targeted for molecular treatment of endometriosis. Furthermore, there is a need for finding out a noninvasive test for endometriosis. Proteomic analysis of HFF, menstrual blood, and

Table 4.1 Comprehensive table with the identified protein biomarkers for different female infertility conditions

Condition	Sample	Methods used	Proteins identified	Reference
Endometriosis	Eutopic endometrium	2D-DIGE and MALDI-TOF	HSP90, annexin A2, peroxiredoxin 2, ribonucleoside diphosphate reductase, prohibitin, hydroxylase, and APOA1	Fowler et al. (2007)
Endometriosis	Eutopic endometrium	2D-DIGE	DJ-1, HSP27, HSP60, HSP70, GRP78, HSP90 beta, MVP, and ERp5	Rai et al. (2010a)
Endometriosis	Frozen endometrial tissue biopsy samples	MALDI-TOF	T-plastin, Annexin V	Kyama et al. (2011)
Endometriosis	Endometrium	LC/MS-MS	Ribosomal proteins like RPL19, RPL5, RPL11, and RPL23, RPL5, RPL11 and RPL17	Xu et al. (2015)
Adenomysis	Ectopic and Eutopic endometrium	2D-DIGE and ESI-Q-TOF	Annexin A2	Zhou et al. (2012)
Adenomysis	Eutopic and ectopic endometrial primary stromal cell	iTRAQ coupled with LC/MS-MS	Vimentin	Marianoswski et al. (2013)
Adenomysis	Ectopic and eutopic endometrium	Nano-LC/MS/MS	HIF1A	Kasvandik et al. (2016)
Endometriosis	Follicular fluid	Nano-UPLC/nano-ESI-MS/MS	Amyloid-b A4 precursor protein-binding family A3, peroxisomal targeting signal 1 receptor, PPR3B and protein FAN, erotransferrin, IGL@ and IGLC1 Protein, IL-2, TAKL protein, pre-B-cell leukemia transcription factor 3 (PBX3) and BMI1 polycomb ring finger oncogene	Lo Turco et al. (2010)

(continued)

Table 4.1 (continued)

Condition	Sample	Methods used	Proteins identified	Reference
Endometriosis	Menstrual blood	2D-DIGE and ESI-Q-TOF/MS	CRMP2, UCH-L1, MYL9	Hwang et al. (2013)
Endometriosis	Plasma	LC/MS-MS	Serotransferrin, 1 complement C3, serum amyloid P-component, alpha-1-antitrypsin and clusterin	Ferrero et al. (2009)
Endometriosis	Plasma	SELDI-TOF-MS	IL-1, IL-8, IL-10, TNFa, RANTES, VEGF, SAA, MCP-1, IL-6, cancer antigen (CA) 125, CA 19.9 and CCR-1	Wolfler et al. (2011)
Endometriosis	Plasma	2-DIGE-LC/MS-MS	Haptoglobin	Hwang et al. (2014)
Endometriosis	Serum	MALDI-TOF/TOF-MS	Haptoglobin (HP), Ig kappa chain C region (IGKC), alpha-1B-glycoprotein (A1BG)	Dutta et al. (2015)
Endometriosis	Urine	LC/MS-MS	Cytokeratin-19	Tokushige et al. (2011)
PCOS	Follicular fluid	LC/MS-MS	Amphiregulin; heparan sulfate proteoglycan 2; tumor necrosis factor, α-induced protein 6; plasminogen; and lymphatic vessel endothelial hyaluronan receptor 1, suprabasin; S100 calcium-binding protein A7; and helicase with zinc finger 2, transcriptional coactivator	Ambekar et al. (2015)
PCOS	Follicular fluid	LC/MS-MS	α1-antitrypsin, apolipoprotein A-I and transferrin	Dai and Lu (2012)
PCOS	Plasma	2D-DIGE and MALDI-TOF	Haptoglobin, alpha-2-macroglobulin, and transferrin and in kappa-free light chain	Insenser et al. (2010)
Unexplained infertility	Endometrium	2D-DIGE and LC/MS	Sorcin, Cofilin-1, Apo-A1 and Ran	Manohar et al. (2014)

even of urine, serum, and peritoneal fluid could be a good alternative method. This would be highly useful for the early detection of endometriosis in symptomatic women who have pelvic pain and/or subfertility with normal ultrasound results.

In the eutopic endometrium of women with endometriosis, Fowler and coworkers found abnormal expression in the following categories of proteins: (i) molecular chaperones, including heat-shock protein 90 and annexin A2; (ii) proteins involved in cellular redox state, such as peroxiredoxin 2; (iii) proteins involved in protein and DNA formation/breakdown, including ribonucleoside diphosphate reductase, prohibitin and prolyl 4-hydroxylase; and (iv) secreted proteins, such as apolipoprotein A1 (Fowler et al. 2007). Kyama's group has proposed T-plastin and annexin V as potential endometrial biomarkers for the diagnosis of minimal to mild endometriosis (Kyama et al. 2011). However, Rai and co-researchers carried out an extensive proteomic analysis of eutopic endometrium from women at different stages of endometriosis (Rai et al. 2010a). They did a comparison of eutopic endometrium in the proliferative phase of menstrual cycle from women having stage IV endometriosis, and the eutopic endometrium in the secretory phase of menstrual cycle from women having stages II, III, or IV endometriosis with women without endometriosis. An interesting stage-specific expression was seen in two isoforms of gelsolin, an actin-binding protein. While the expression of gelsolin isoform 'a' decreases, the expression of gelsolin isoform 'b' increases from stage II to stage IV endometriosis. This deregulation in gelsolin expression was assumed to be associated with facilitating malignant transformation, invasion, and metastasis. The differentially expressed protein profile of each patient category was dominated by structural proteins. Several proteins associated with functions such as stress response, protein folding, protein turnover, RNA biogenesis, protein biosynthesis, nuclear proteins, and energy metabolism are seen to be altered in eutopic endometrium of endometriotic women (Rai et al. 2010a). Proteins related with oxidative stress such as peroxiredoxin 2 isoform 'a' and peroxiredoxin 3, and proteins playing a role in immunity such as annexin A4 and annexin A5, have been found to be downregulated in endometriosis. On the other hand, proteins involved in cell cycle, cell proliferation, and anti-apoptosis, namely 14-3-3 protein epsilon, are found to be upregulated in endometriosis. Recently, Xu and colleagues carried out the phosphorylation profiling of endometrium from endometriosis patients and identified 516 proteins that were modified at the phosphorylation level during endometriosis. Further pathway analysis indicated that the ribosome pathway and focal adhesion pathway were the top two pathways altered in females with endometriosis (Xu et al. 2015).

Adenomysis is uterine endometriosis, characterized by downward extension of the endometrium into the uterine myometrium. Analysis of the differentially expressed proteins in matched ectopic and eutopic endometrium of adenomyosis patients revealed that a group of estrogen-responsive proteins was found significantly upregulated in the ectopic endometrium of adenomyosis compared with its eutopic counterpart (Zhou et al. 2012). The deregulated proteins were mostly cytoskeletal (27 %), while others were involved in signal transduction, redox regulation, proliferation, and apoptosis. Furthermore, overexpression of annexin A2 was marked in ectopic lesions of human adenomyosis and was found to be associated with markers of epithelial to mesenchymal transition and dysmenorrhea severity of adenomyosis patients. It also enhanced both the metastatic potential and pro-angiogenic capacity of adenomyotic endometrial cells. In ovarian endometrioma, cytoskeletal protein vimentin was found to be significantly increased, which is suggestive of its role as a potential biomarker (Marianowski et al. 2013). The overexpression of vimentin correlates well with accelerated tumor growth and invasion. In a recent study, the proteome of ectopic endometriotic stromal cells from peritoneal endometriotic lesions was compared with that of eutopic endometrial stromal cells of endometriosis patients and fertile controls (Kasvandik et al. 2016). Significant anomalies with metabolic reprogramming were observed in ectopic stromal cells with extensive upregulation of glycolysis and downregulation of oxidative respiration. Additionally, pathogenesis of endometriosis was accompanied by attenuated apoptotic potential, increased cellular invasiveness and adhesiveness, and altered immune function in ectopic endometriotic stromal cells.

In addition to tissues analysis, follicular fluid is also an option that could be used to find biomarkers of endometriosis. A case-control study involving women with Stages III or IV endometriosis and pregnant women without endometriosis, both submitted to controlled ovarian stimulation for in vitro fertilization, identified at least 64 differentially expressed proteins, which may be related to the physiopathology of endometriosis (Lo Turco et al. 2010). Further, another study by the same group demonstrated a differential expression of proteins in endometriotic patients achieving pregnancy and those not achieving pregnancy by ART (Lo Turco et al. 2013). Mostly the proteins responsive to oxidative stress and apoptosis are deregulated in the latter group. They found that certain proteins, namely, apolipoprotein-AIV, transthyretin, complement factor I, vitronectin, kininogen-1, and FAK1 have great potential to become biomarkers and subsequent therapeutic targets for endometriosis. A functional annotation of HFF proteins found in the endometriosis group revealed that the most prevalent proteins were involved in transport, binding, as well as regulation and metabolism of steroids, emphasizing the sex hormone-binding globulin (Regiani et al. 2015).

Menstrual blood can also be taken as a noninvasive and quick diagnostic tool for detection of endometriosis. A proteomic analysis of eutopic endometrial cells collected from the menstrual blood of females with or without endometriosis was performed to identify novel potential biomarkers for endometriosis (Hwang et al. 2013).Three proteins in menstrual blood, namely, collapsin response mediator protein 2 (CRMP2), ubiquitin carboxyl-terminal hydrolase isozyme L1 (UCH-L1), and myosin regulatory light polypeptide 9 (MYL9), were suggested as candidate biomarkers of endometriosis.

Other biological fluids can also be relevant in providing the list of proteins having aberrant expression and playing a key role in the pathogenesis of endometriosis. Several studies have carried out a protein profiling of peritoneal fluid (PF) of women with endometriosis (Ferrero et al. 2007, 2009; Wolfler et al. 2011). Proteins such as serotransferrin, complement C3, serum amyloid P-component, alpha-1-antitrypsin, and clusterin were upregulated in PF of the patient group (Ferrero et al. 2009). Similarly, proteomic analysis of plasma samples obtained during the menstrual phase can also enable the diagnosis of endometriosis that is undetectable by ultrasonography, with high sensitivity and specificity (Hwang et al. 2014; Fassbender et al. 2012). Moreover, evaluation of novel serum biomarkers in women with endometriosis showed that for diagnosis of Stage I, proteins such as Ig kappa chain C region (IGKC) and haptoglobin (HP) may be proposed, while HP, IGKC, and alpha-1B-glycoprotein (A1BG) can be considered as effective serum protein markers for the diagnosis of Stages II, III, and IV endometriosis (Dutta et al. 2015). Investigation of the altered urine proteome in women with endometriosis has also demonstrated a cohort of differentially expressed proteins (Wang et al. 2014b; Tokushige et al. 2011). Protein cytokeratin-19 was proposed to be an efficient urinary biomarker for endometriosis in one of the studies (Tokushige et al. 2011).

4.8 Polycystic Ovary Syndrome

Polycystic ovary syndrome (PCOS) is a common gynecological and endocrinological disease, associated with excessive secretion of androgens. Although characterized to be a heterogeneous disorder having multiple short and long-term consequences, yet the etiology remains unclear. Women with PCOS are known to have fertility and reproductive problems, along with being more susceptible to several metabolic disorders such as diabetes mellitus, metabolic syndrome, obesity, and cardiovascular disease. PCOS is thought to be the result of the interaction between predisposing genetic variants with environmental factors, including diet and lifestyle. The current diagnosis of PCOS involves clinical identification of at least two of the following three criteria: chronic oligo-/anovulation, clinical and/or biochemical signs of hyperandrogenism, and ultrasound evidence of polycystic ovaries. Proteomic analysis would help us to unravel the molecular factors behind the pathogenesis of PCOS and determine the risk factors associated with oocyte arrest, other metabolic disorders, spontaneous abortions and complications in pregnancy in these patients.

Follicular fluid is one of the major reproductive fluids used for proteomic analysis of PCOS patients to find the effect of PCOS on folliculogenesis. A comparative protein profiling of HFF collected from PCOS and normo-ovulatory women undergoing ART identified a list of about 186 proteins showing differential abundance between controls and women with PCOS (Ambekar et al. 2015). Proteins involved in various processes of follicular development including amphiregulin; heparan sulfate proteoglycan 2; tumor necrosis factor, α-induced protein 6; plasminogen; and lymphatic vessel endothelial hyaluronan receptor 1 were found to be deregulated in PCOS. Another study correlated the abnormal expression of HFF proteins involved in lipid metabolism, to the PCOS patients' inability to ovulate normally (Dai and Lu 2012). Further, an upregulation of alpha-1-antitrypsin in PCOS patients' HFF might be related to the failure of oocyte maturation in PCOS. Serum and plasma are the other extensively studied biological fluids in PCOS patients (Zhao et al. 2008; Insenser et al. 2010, 2013). Based on analysis of the proteome database of

PCOS women, the involvement of several pathways in the pathophysiology of PCOS was derived, with the major ones being metabolic pathways, inflammation, and oxidative stress (Insenser and Escobar-Morreale 2011; Atiomo et al. 2009).

Systematic reviews targeting to identify biomarkers for certain diseases in women with PCOS have also been carried out by integrating proteomic biomarkers for the disease with previously published databases of all proteomic biomarkers identified so far in PCOS women. Galazis' group found nine differentially expressed protein biomarkers in women with endometrial cancer and PCOS, which includes transgelin, pyruvate kinase M1/M2, gelsolin-like capping protein (macrophage capping protein), glutathione S-transferase P, leucine aminopeptidase (cytosol aminopeptidase), peptidyl-prolyl cis–transisomerase, cyclophilin A, complement component C4A, and manganese-superoxide dismutase (Galazis et al. 2013). Previously, the same group have documented a panel of six possible biomarkers for ovarian cancer in women with PCOS, which included the proteins calreticulin, fibrinogen-γ, superoxide dismutase, vimentin, malate dehydrogenase, and lamin B2 (Galazis et al. 2012). Recently, five proteomic biomarkers were demonstrated to be differentially expressed in women with pre-eclampsia and PCOS, namely, transferrin, fibrinogen α, β, and γ chain variants, kininogen-1, annexin 2, and peroxiredoxin 2 (Khan et al. 2015).

4.9 Unexplained Infertility

Unexplained infertility is a diagnosis that is made by negative tests for known diseases that cause female infertility and also by ruling out the male factor. Thus, it would be helpful to have proteins that indicate a case of unexplained infertility. The embryo–uterus interaction is modulated by a wide range of molecules. Any alteration in these molecules can lead to implantation failure, pregnancy loss or infertility (Cakmak and Taylor 2011). Compromised endometrium receptivity is a central cause of unexplained infertility in women (Manohar et al. 2014).

To our knowledge, the number of studies involving proteomics in women with unexplained infertility is limited. Manohar et al. compared the endometrium of women with unexplained infertility during early (LH +2) and mid-secretory (LH +7) phases of the cycle (Manohar et al. 2014). Proteasome subunit alpha type-5 (PSMA5), involved in protein degradation, is seen to be downregulated during the mid-secretory phase, which may lead to impaired differentiation of the endometrial cells due to improper regulation of apoptosis. They also found a decreased level of superoxide dismutase during the mid-secretory phase, showing that a dysfunctional antioxidant system might, once more, contribute toward unexplained infertility. On the other hand, apoliprotein, an anti-inflammatory protein that inhibits the synthesis of adhesion molecules which are very likely important for embryo implantation, was found to be upregulated. Upregulation of RAN GTP-binding nuclear protein (Ran) in the mid-secretory phase can be correlated with the altered development and progression of endometrial stromal cells.

Chapter 5
Proteomics in Assisted Reproduction

Luna Samanta

Ever since the birth of the first in vitro fertilization (IVF) baby in 1978, the assisted reproductive technologies (ART) has seen a bloom and have enabled millions of people worldwide to have biological children who otherwise would not have been able to do so (Andersen et al. 2008; de Mouzon et al. 2010; Ferraretti et al. 2012). Nevertheless, majority of embryos (~ 70 %) fail to implant and a few lead to successful pregnancy (Coughlan et al. 2014). The identification and transfer of the most viable embryo is one of the deciding factors for maximal pregnancy rates. The method of choice to assess embryo quality in all ART clinics is based upon morphological scoring, such as cell shape, cell cycle timing, and compaction times (Ziebe et al. 1997), including continuous monitoring using time-lapse systems (Massip and Mulnard 1980). However, morphological assessment to determine embryo potential is universally believed to be subjective with limited potential in distinguishing competent embryos (Machtinger and Racowsky 2013). Another important difficulty in ART is multiple pregnancies leading to increased risks of serious complications for the mother and babies and the impact on healthcare costs (Thurin et al. 2004). Thus, single embryo transfer is gaining momentum across the ART clinics which further makes embryo selection inevitable for transfer. Therefore, the need of the hour is the identification of reliable noninvasive biomarkers of embryo development for improvements in the efficiency of ART treatment eventually reducing emotional and financial stress of the patients. In this

L. Samanta (⊠)
Department of Zoology, Ravenshaw University, Cuttack, India
e-mail: lsamanta@ravenshawuniversity.ac.in

© The Author(s) 2016
A. Agarwal et al., *Proteomics in Human Reproduction*,
SpringerBriefs in Reproductive Biology, DOI 10.1007/978-3-319-48418-1_5

regard, Omics platforms offer promising outcomes, particularly proteomics as the secretome (proteins secreted by the embryo into the culture medium) for noninvasive assessment of the quality of the embryo could improve success rates in clinical embryology.

5.1 Proteomics of Embryos: Lessons from Animal Studies

Albeit the central dogma of molecular biology states that proteins translated from RNA transcripts are directly responsible for cellular function, yet analysis of transcript level does not correlate with protein abundance. In fact, Williamson et al. have reported low correlation between paired comparisons of changes in mRNA and protein expression levels in embryonic stem cells (Williamson et al. 2008). It is believed that cells utilize several mechanisms such as targeted mRNA degradation and protein degradation to regulate transcription/translation leading to lack of correlation between gene expression, RNA transcripts and protein expression. Therefore, it is logical to have an in depth investigation of the mammalian proteome in order to understand cellular function and comprehend biological processes and/or disease states.

Two-dimensional gel electrophoresis (2-DGE) was used to construct and analyze protein databases of the preimplantation mouse embryo (Latham et al. 1992; Shi et al. 1994) while the expression of known glucose transporter proteins was identified in rabbits (Navarrete Santos et al. 2004). Subsequently, post-translational modifications, such as phosphorylation, in relation to embryo development were also reported (Wang et al. 2005). With the advent of high throughput mass spectrometry (MS), identification of groups of proteins within limited amounts of complex biological fluids and tissues were possible (Gutstein et al. 2008; Jansen et al. 2008).

5.2 Noninvasive Proteome Prediction of Human Embryo

The metabolism of embryonic cell changes according to the stages of embryonic development. These points toward stage-specific protein expression, of which few proteins and peptides are secreted into the surrounding culture medium contributing to the embryo's secretome.

Presence of soluble human leukocyte antigen G (sHLA-G) in spent IVF medium of Day 3 embryos was correlated with higher pregnancy rates (Noci et al. 2005; Sher et al. 2005). Nevertheless, pregnancies established from sHLA-G-negative embryos are attributed to inability in measuring sHLA-G production in the spent medium (Sargent et al. 2007). O'Neill reported a soluble factor 1-o-alkyl-2-acetyl-sn-glycero-3-phosphocholine (paf) secreted by mammalian preimplantation embryos that acts as a survival factor during embryonic development by acting in an autocrine manner (O'Neill 2005). Since an increase in HLA-G mRNA and protein during preimplantation development period of day 3 onwards is recorded, its measurement on day 5–6 of culture is suggested to be more fruitful (Yao et al. 2005). However, a meta-analysis study by Vercammen et al. indicate the necessity of further research involving single embryo culture, single embryo transfer and more sensitive HLA-G detection techniques so as to establish the accuracy of sHLA-G for predicting pregnancy among women undergoing IVF (Vercammen et al. 2008). On the other hand, the effects of sHLA-G are more apparent in ICSI (Warner et al. 2008).

Another protein, Acrogranin, detected in surrounding media of preimplantation mouse embryos, if added to the culture medium, promotes blastocyst formation, while addition of anti-acrogranin antibodies delayed its formation. Acrogranin is believed to be an autocrine factor that acts directly on the trophectoderm (Diaz-Cueto et al. 2000).

While studying the interactions between the embryo and endometrial epithelial cells in the human in vitro model, Gonzalez et al. observed the small pleiotrophic peptide hormone leptin in the conditioned medium from blastocysts (Gonzalez et al. 2000). A substantially high concentration of leptin is secreted into the conditioned medium of more competent blastocysts than the arrested embryos. It has been hypothesized that the leptin secreted by the blastocyst triggers molecular dialog with maternal endometrium via receptors whose expression increases in the human endometrium during the luteal phase (Cervero et al. 2005).

Further, it is demonstrated that cell-free spent media of embryo grown to blastocyst stage modulate the expression of HOXA10 in Ishikawa cells (epithelial endometrial cell line). This suggests its role in embryo-endometrial interaction, which could transform the local uterine environment, impacting both embryo development and the implantation process (Sakkas et al. 2003). Recently, Mains et al. correlated increased levels of Apolipoprotein A1 to higher morphological grade of blastocyst (Mains et al. 2011). Moreover, apoptosis inhibitor, Survivin is expressed and secreted by oocytes and embryos and plays an important role in embryogenesis (Balakier et al. 2013). Proteins, such as granulocyte macrophage colony-stimulating factor (GM-CSF) and chemokine ligand 13 (CXCL13), were found to be lower in the secretome of the implanted human blastocysts (Dominguez et al. 2008).

In most of the studies, only a single factor was studied for assessing the competence of the embryo, however, embryo development involves more complex pathways. Therefore, multiple secretome markers are necessary to be identified and characterized to predict developmental competence and/or implantation potential of the embryo. Recently, a total of 15 proteins belonging to the positive implantation group were identified and the most representative is Jumonji protein (JARID2), regulator of histone methyltransferase complexes that plays an essential role in embryonic development (Cortezzi et al. 2011).

5.3 Proteome Biomarkers of the Embryo

Katz-Jaffe et al. reported that the morphological development of human blastocyst is associated with significant alterations in the expression of proteins (Katz-Jaffe et al. 2006). Thus, identification of proteomic biomarkers will not only help us in understanding the mechanisms of the biological processes that occur at the cellular level during preimplantation embryonic development, but also help in quantification of embryonic viability potential to increase IVF pregnancy rates and live births while reducing the number of transferred embryos (Katz-Jaffe et al. 2006). In this regard, proteins involved in apoptotic and growth-inhibiting pathways may be pivotal candidates to be closely involved in this process (Katz-Jaffe et al. 2006). Recently, Poli et al. for the first time characterized and quantified blastocoel proteins secreted by single human embryos prior to implantation, using a minute sample volume (5 μl) to predict embryo competence (Poli et al. 2015).

5.4 Endometrium

It is not only the reproductive competence of blastocysts, but also the preparedness of the endometrium to accept the embryo that is equally important in the process of implantation. Proteomics can serve to develop receptivity markers of endometrium as any communication between two tissues is established via protein-protein interaction. Since the endometrium undergoes characteristic changes during different phases of the menstrual cycle, several studies are focused on the identification of the proteome during these phases. Stathmin 1 and annexin A2 are differentially regulated in receptive and nonreceptive endometrium while Annexin 4 plays a crucial role in the receptive process (Dominguez et al. 2009). When the proteome of proliferative and secretory phase of endometrium was compared, 194 proteins were detected, of which seven were differentially expressed (Rai et al. 2010b). Comparative proteomics of pre-receptive (day 2 after LH surge) and receptive (day 7 after LH surge) phases reveal 31 proteins to be supposedly involved in the implantation process (Li et al. 2011b). On the other hand, mid-secretory (receptive) phase is characterized by downregulation of calecticulin, fibrinogen adenylatelinase isoenzyme-5, and transferrin together with and up regulation of annexin 5, alpha-1-atitypsin, peroxiding-6, and creatine kinase (Upadhyay et al. 2013).

During decidualization of endometrium, Garrido-Gomez et al. identified 60 differentially expressed proteins in endometrial stromal cells (ESCs) after in vitro decidualization when compared to control ESCs, including 36 over-expressed and 24 under-expressed proteins. Cathepsin B, tranglutaminase 2, peroxiredoxin 4, and (cytosolic beta-actin were identified as markers of decidualization by the authors (Garrido-Gomez et al. 2011). They also reported 11 secreted proteins to be up-regulated and 2 to be down-regulated, of which IGF binding protein-1, prolactin, myeloid progenitor inhibitory factor-1, and platelet endothelial cell adhesion molecule-1 were identified as markers (Garrido-Gomez et al. 2011). In another study comparing the proteome profile of the receptive to nonreceptive endometrium, the authors found progesterone receptor membrane component 1 (PGRMC1) and annexin A6 (ANXA6) proteins to be targeted for the evaluation of endometrial receptivity (Garrido-Gomez et al. 2014).

Recurrent implantation failure (RIF) is believed to be associated with a characteristic protein profile different from fertile women. Apolipoprotein A-I is predicted as an endometrial anti-implantation factor (Brosens et al. 2010). This protein was highly expressed in mid-secretory eutopic endometrial tissue from patients with endometriosis, suggesting its dysregulation as a cause of RIF. In fact, endometriosis is characterized by differential protein profile including the downregulation of heat shock protein 90-alpha and beta, implying the role of comparative proteomics in disease diagnosis and/or prediction of pregnancy outcome (Fowler et al. 2007).

Chapter 6
Challenges of Proteomic Studies in Human Reproduction

Gayatri Mohanty and Luna Samanta

The human reproductive system is a set of interconnected processes (Wang et al. 2003), which involves the direct interaction of the sperm and the oocyte, fusion of cell membranes and the subsequent union of maternal and paternal genomes (Primakoff and Myles 2002). Biological systems consist of several cellular entities which include RNAs, metabolites and proteins that participate in a tightly regulated manner though intricate orchestration of cellular processes. With the completion of the human genome project a decade ago, a large set of readily accessible database have been yielded, which helps in the understanding of the functional ability of the tissue or specific cells at the molecular level, both in a healthy and diseased state. However, DNA sequence data alone cannot answer questions about the level of protein expression, so the emphasis is shifting to the protein complement of the human organism. The question then arises as to why biochemists should place an emphasis on reproductive biology? The answer to this is obvious as half of the world seems to suffer from problems associated with human reproduction while the other half seems to have been afflicted with the inability to reproduce—a disorder termed as infertility.

G. Mohanty (✉) · L. Samanta
Department of Zoology, Ravenshaw University, Cuttack, India
e-mail: gayatri_mohanty32@yahoo.com

L. Samanta
e-mail: lsamanta@ravenshawuniversity.ac.in

© The Author(s) 2016
A. Agarwal et al., *Proteomics in Human Reproduction*,
SpringerBriefs in Reproductive Biology, DOI 10.1007/978-3-319-48418-1_6

Assisted reproductive techniques are becoming an indispensable part of fertility treatment in patients with acute reproductive failure. The establishment of non-invasive, reliable means of assessing endometrial receptivity, embryonic viability, aneuploidy and fertilizing ability of the spermatozoa is the need of the hour (Gupta et al. 2015). Proteins form the core of every reproductive process and their altered behavior have been implicated in a myriad of pathological conditions and thus by a major class of drug targets. While the genome of a cell is static, the proteome of that same cell is modulated by varying environmental conditions and developmental cues and is a plausible etiology for several reproductive failures. It is therefore becoming increasingly clear that in order to truly understand a protein, we need to identify the entire protein complement, detect covalent modifications and allow for quantitative comparisons between samples. In contrast with some biochemical methods and allied technologies that are restricted to the analysis of only a single protein in a cell or tissue (Wilkins et al. 1996), proteomics deals with the analysis of the entire population of expressed proteins, based on their expression, localization, functions, post-translational modifications and interactions at a specific condition and time (Gupta et al. 2015).

Proteomics has emerged as a frontline in the field of reproductive biology research and has been extensively used to investigate protein alterations in different diseased states (Agarwal et al. 2015b). Proteins are a group of biomolecules and its assemblage represents the mystifying processes in molecular life sciences. Any smallest hitch in its assembly into functional macromolecular complexes may have a profound effect on its cellular activity (Hartl and Hayer-Hartl 2002, 2009). Thus, the question of how protein activity is regulated as protein binding requires an integration of structural, functional and dynamic information is a query that needs to be answered. Of late, proteomic studies have not only been used to generate a repertoire of proteins in a particular physiological state but also to screen for differentially expressed proteins in patients afflicted with various etiologies in relation to human reproduction. Several pathologies are the result of improper protein folding that hampers the binding of their cofactor and subsequently leads to non-specific protein aggregation (Ross and Poirier 2004).

Protein separation techniques have been greatly improvised in the last decade which includes two-dimensional polyacrylamide gel electrophoresis (2D-PAGE), liquid chromatography (LC), isotope-coded affinity tag (ICAT) labeling, and so on. In addition, protein analysis from tryptic digests using mass spectrometry (MS) technology and database searching has achieved much progress. Although success of proteomic analysis in human reproduction have been well documented, there are several cases where the technique is still in its infancy in solving the many unanswered questions of human reproduction. This chapter is not meant to place proteomic analysis and its relationship with human reproduction in a negative perspective, rather to focus on some of the important issues that have been relatively less attended.

6.1 Goals and Challenges

The standard method for the study of any cell, tissue or biological fluid is by separating the proteins on a 2D-PAGE gel and identifying the separated proteins by MS (either peptide mass fingerprinting or tandem mass spectrometry analysis) (Giannopoulou et al. 2009). Most proteomic studies today still follow this approach. However, there lie several limitations to this methodology. With the help of two-dimensional difference gel electrophoresis (2D-DIGE), only 30–50 % of the entire proteome is able to be visualized. The low abundant proteins, as well as some proteins that cannot be separated on a 2D-PAGE gel due to their physicochemical properties (isoelectric point, hydrophobicity, molecular weight), remain undetected. This results in the complete negligence of membrane proteins and small or low abundant proteins or peptides (Chandramouli and Qian 2009). New innovative strategies for the enrichment of less abundant proteins are required in order to facilitate the efficient discovery of 'low abundance' biomarkers. To overcome some of these limitations, several gel-free high-throughput technologies for proteome analysis have been developed. The recent development of bottom-up or shotgun proteomics has opened vast avenues in the identification and characterization of cellular proteome. Wherein this technology, proteins are cleaved into peptides using proteolytic enzymes and subsequently these peptides are separated, usually by liquid chromatography, and subjected to tandem mass spectrometric analysis. Mass spectrometric identification of these peptides allows the determination of the protein content of the initial sample.

6.2 Sample Complexity

A significant amount of challenges are encountered while working with various reproductive samples, which includes tissues, fluids and related biological materials. From an andrological point of view, a major limitation in the analysis of the entire testis lies in the existence of a mixed cellular population. As an alternative, the potential development of efficient spermatogenic in vitro culture systems allows for the synchrony of relatively pure cellular stages of a spermatozoa's differentiation mechanism. Due to their low or non-invasive mode of collection, biological fluids such as plasma or urine are favored in most clinical applications. However, these are not an ideal choice for proteomic studies because of its vast dynamic range (at least 6–12 orders of magnitude) and dominance of highly abundant plasma proteins that makes the analysis and identification of low molecular mass proteins from plasma difficult (Anderson 2010). This limitation is similarly applicable in the case of a single spermatozoon, which apparently turns out to be a complex mixture of different cell types and sperm subpopulations with substantial biochemical, functional and morphological differences. Reproductive tissues—in particular, endometrial tissue —are also inherently heterogenous with respect to developmental, temporal and biological composition. Fractionation is therefore an essential component for any analysis of biological sample (Feist and Hummon 2015). This fact together with the variation present in independent individuals and the physiological changes that occur in the sample which is being studied must be taken into consideration. For example, upon ejaculation, the spermatozoon generates an enormous potential for variation, which must be taken into account in proteomic studies (Oliva and Castillo 2011). Proteins are an integral part of every cellular event and hence understanding their molecular nature is of utmost importance. Proteins, in being the key players in the many cellular networks, add several difficulties in understanding their physical and biochemical nature, which is not inherent in the study of nucleic acids. Structurally, proteins can be categorized into secondary and tertiary structures and notably these structures are essential for protein function. A variety of biochemical properties are exhibited by proteins that far exceeds the relatively homogenous behavior of oligonucleotides and is crucially dependent on the precise three dimensional (3D) structure of the folded polypeptides. Proteins are easily denatured by the action of enzymes, heat, light or by aggressive mixing, as in beating egg whites (Cho 2007). Sometimes, the poor solubility of proteins makes it difficult to analyze the molecular nature of proteins. On the other hand, from a molecular point of view, proteins cannot be amplified as performed in DNA samples, therefore it is often difficult to detect less abundant species in any given protein sample. Moreover, unlike RNA or DNA, proteins do not inherently possess well-defined high-affinity and/or high-selectivity binding partners (Aebersold and Cravatt 2002). Another great challenge faced by the proteomic technology today is the inherent complex nature of cellular proteome, which is attributed to many factors. Firstly, the number of proteins in a cellular proteome is vast, although no definite calculations can be carried out. It has been estimated that up to 50,000 protein species can be simultaneously present in a eukaryotic cell. In addition, there is a dynamic range of protein expression (which spans seven or eight orders of magnitude) that is present in vastly different quantities. Furthermore, the structural diversity along with various physiochemical characteristics of the proteins present in a cellular proteome contributes further to its complexity.

6.3   Mass Spectrometry Analysis

Mass spectrometry analysis is a booming technology in the field of reproductive sciences, as it offers a new tool to trace various compounds and metabolites. Basically, a mass spectrometer is composed of three components: the ion source, the mass analyzer and the ion detection system (Scherl 2015). Yet as with many other technologies, there is always a risk factor associated that will be difficult to meet because of technological hindrances—some of which are discussed below. Proteins and peptides being polar, non-volatile and thermally-unstable require an ionization technique that transfers an analyte into the gas phase without extensive degradation. The introduction of soft ionization methods is one of the important developments in instrumentation that has augmented the analysis of proteins and peptides by MS (Yates et al. 2009). There are four ionization methods that are currently utilized in proteomic experiments—electrospray ionization (ESI), matrix-assisted laser desorption/ionization (MALDI), surface-enhanced laser desorption/ionization (SELDI), and desorption electrospray ionization (DESI). Both MALDI and ESI are effectively used for proteomic studies coupled with tandem mass spectrometry (MS/MS) analysis. MALDI generates singly charged ions and is effective for peptides and proteins beyond 100 kDa. The ionization process in MALDI requires several hundred laser shots to achieve an acceptable signal-to-noise ratio for ion detection. ESI generates multiply-charged ions that are useful in the measurement of protein structure and folding, as well as for high-throughput protein identification (Peng et al. 2008). The drawbacks in MALDI are low shot-to-shot reproducibility and strong dependence on sample preparation methods (Zheng et al. 2005). As an alternative, recently, efforts have been made to develop new methods, as a variation to ESI and MALDI, which includes DESI and SELDI, which offer potential enhancements in sensitivity and ease of sample introduction.

Mass analyzers form an integral component of each instrument because they can store ions and separate them based on the mass-to-charge ratios. There are various types of mass analyzers currently used in proteomic research. Based on its differences as to how the mass-to-charge ratio (m/z ratio) of peptides is determined, the efficiency of the instrument can be analyzed.

A critical number of elements exist in the workflow of MS analysis that is known to strongly affect the outcome of an experiment. Sample preparation, ion suppression, sensitivity and data analysis are some of the key features of a MS analysis that needs critical evaluation. The term ion-suppression refers to the state in which surface ionization processes do not result in identical ionization efficiencies for each chemical species on the surface (Baggerman et al. 2005). A high incidence of co-elution is a common phenomenon in complex samples and poses a limiting factor for the number of peptides to be analyzed in an LC-MS/MS analysis. This nature of the complex analyte being studied makes it nearly impossible to generate quantitative mass spectrometric images without evaluating the ion suppression effects. A chromatographic peak is only available for a short time frame in which MS, as well as several MS/MS spectra, have to be acquired. Co-elution limits the number of ions that can be selected for fragmentation as very abundant peptides suppresses the signal of the less abundant ones, leading to ion-suppression (Mirza and Olivier 2008). The ion-suppression effect is a major hindrance for quantitation in many proteomic analyses. Considering the potential impact on the protein quantitation, sample preparation protocols should be so designed which should involve pre-fractionating very complex samples and thus reducing the complexity of each fraction so as to limit the effect of ion-suppression (Heeren et al. 2009). Thus, the shotgun proteomics approach combines the high separation efficiency of multidimensional chromatography with the powerful peptide identification capacity of ESI tandem mass spectrometry. This multidimensional system approach (also known as MudPit), though having highly sophisticated technology, has a number of limitations. First and foremost, the acquisition of thousands of MS/MS spectra poses a problem from the aspect of storage and data analysis. In addition, peptide samples obtained from enzyme digests are complex and only a fraction is randomly selected for fragmentation as they elute from the column. This results in low reproducibility of the data and repeated analyses causes variability in the list of identified proteins (Baggerman et al. 2005).

An alternative approach to bottom-up proteomics is the top-down method wherein the intact protein is directly dissected and amino acid sequence information is obtained by dissociation. Like all other modern techniques, this is also associated with many obstacles and limitations. This approach results in the formation of multiply-charged protein precursor ions and hence poses a difficulty in the interpretation of the top-down fragmentation spectra. There are several means to overcome this limitation. This is possible by reducing the charged states on the product ion through the introduction of gas-phase anions to strip protons from the product ions through ion-ion proton transfer reaction. The limitation can also be overcome through the use of either a Fourier transform ion cyclotron resonance (FT-ICR) mass spectrometer or Orbitrap mass spectrometer with high mass accuracy.

6.4 Protein Quantitation

An LC-MS/MS experiment has the capacity to identify and quantify thousands of proteins in complex mixtures. Beyond identification, quantitation of analytes can also be performed by looking at the relative intensity response from the mass spectrometer. Traditionally, protein quantification is carried out with gel-based approaches. An extension of quantitation is differential analysis, in which two or more samples are compared for discriminating features. The analysis of differential protein expression in complex biological samples requires strategies for rapid, highly reproducible and accurate protein quantitation (Mirza and Olivier 2008). Differential expression analysis can be used as a part of a biomarker discovery, with the hope that the features that are found to be discriminating between groups will have clinical utility. However, the workflow has a number of deficiencies. In contrast to the gel-based approach, stable isotope labeling and label-free methods have been established that involve mass spectral analyses and result in quantification at the peptide level. Accordingly, some of the most commonly applied approaches for quantitation and identification of biomarkers include the use of stable isotope labeling, which can be classified into metabolic labeling, chemical mass tagging, and enzymatic labeling. To overcome the limitations of gel-to-gel reproducibility in the two dimensional electrophoresis method, gel-based quantification approaches such as 2D-DIGE has been established. Proteins are fluorescently labeled with cyanine dyes such as Cy2, Cy3, or Cy5 prior to 2-DE. It is advantageous in the context that proteins in each of up to 3 samples can be labeled with one of these fluorescent dyes, and the differentially labeled samples can be mixed and loaded together on one single gel, allowing the quantitative comparative analysis of three samples using a single gel. Although the technique is advantageous in the quantitative analysis of proteins and identifying post-translational modifications, it has limitations in resolving hydrophobic proteins, interference of high abundance proteins and poor resolution of spots. Furthermore, proteins without lysine cannot be labeled, and they require special equipment for visualization, and fluorophores are very expensive (Chandramouli and Qian 2009). Finally, multiple protein isoforms can often be found in different spots on the gel, complicating the comprehensive analysis. Alternatively, gel-free based proteomic technique is emerging as the method of choice for quantitatively comparing protein levels among biological proteomes, since they are more sensitive and reproducible compared to the two-dimensional gel-based methods. In contrast to the gel-based approach, the stable isotope labeling method involves labeling of proteins/peptides with one or more stable isotopes and pooling the samples with an unlabeled control sample (Ong et al. 2002). Metabolic isotope labeling is a one of its kind approach that requires in vivo incorporation of isotope-labeled essential amino acids during cell growth, which is termed as stable isotope labeling by amino acids in cell culture (SILAC) (Aebersold and Mann 2003). SILAC as a technique involves in vivo coding that requires no chemical manipulation, and there lies very little chemical difference between the isotopically-labeled amino acid and its naturally occurring counterpart (Amanchy et al. 2005). Although the technology is quite promising, it is limited to studies that involve cell culture so that cells can incorporate the exogenous amino acid into proteins (Ong et al. 2003). Under such circumstances, the study of reproductive tissue samples is precluded. Chemical mass tagging involves proteins or peptides that are tagged with a stable isotope-containing molecule. The isotope-coded affinity tag (ICAT) is perhaps the best-known of these reagents for stable isotope labeling and serves as a good paradigm for understanding similar reagents. The ICAT reagent consists of a sulfhydryl-reactive iodoacetate group, which is a linker carrying light or heavy isotopes and a biotin affinity group that facilitates peptide enrichment (Sethuraman et al. 2004). The ICAT experiment involves the labeling of protein samples with either light or heavy reagents on cysteine thiols. The mixtures of labeled proteins are then digested by trypsin and separated through a multistep chromatographic separation procedure. The specificity of the ICAT reagent to cysteines enables the reduction in the complexity of the original peptide mixture to a few tryptic peptides. This is followed by the identification of peptides with tandem MS, and the relative quantifications of peptides are inferred from the integrated LC peak areas of the heavy and light versions of the ICAT-labeled peptides. However, this technology is not universally acceptable, although it works well in narrowing down the dynamic range of the proteome due to the enrichment of only tagged peptides. As a result by means of ICAT, only a fraction of peptides will be analyzed, while ICAT is silent with regard to the significant number of proteins lacking cysteine residues (von Haller et al. 2003). The technique is also blinded to any non-cysteinilated peptide as well, making it ill-suited for the analysis of PTMs or protein isoforms generated by alternative mRNA splicing. In recent years, additional affinity tags have been developed that allow specific binding to other amino acid residues, but it still involves an enrichment step and untagged proteins or peptides will not be analyzed and identified. It has also been noted that the ICAT tag is a relatively large molecule compared to the small peptides it labels, which may result in interference with peptide ionization and inaccurate mass spectra (Yan et al. 2004). Recently, isobaric tags for relative and absolute quantification iTRAQ have been established wherein isobaric mass tags are used to label N-termini and lysine side chains of peptides prior to fractionation (Aggarwal et al. 2006). The iTRAQ reagent is well known for carrying out the relative and absolute quantitation of proteins. The reagent consists of an amine-specific reactive group, a mass balance group, and a reporter group. The reporter and balance groups carry stable isotopes, with different combinations of isotopes in the reporter group but with uniform molecular weight in the combined molecule. After co-elution, the peptides are further fragmented to release the iTRAQ reagent which enables for generating both protein identities and quantitative information by MS in a single experiment (Chong et al. 2006). The iTRAQ technology offers several advantages, which include the ability to multiplex several samples, i.e. a maximum of eight different samples may be combined prior to analysis, along with quantification, simplified analysis and increased analytical precision and accuracy. The limitation in the technique lies in the fact that reporter ion masses are below the low-mass cutoff of ion-trap mass spectrometers, thus requiring more advanced mass spectrometers for the analysis (Mirza and Olivier 2008). Enzymatic labeling is another commonly used method of stable isotope labeling, which involves enzymatic incorporation of ^{18}O atoms during proteolytic cleavage, most commonly by trypsin, which results in peptides with either one or two ^{18}O atoms at the carboxy terminus. The ^{18}O labeling is more advantageous than the aforementioned techniques in that, as in ICAT, this technique does not favor peptides containing certain amino acids (e.g. cysteine), nor does it require an additional affinity for the enrichment of peptides. It is also applicable to several biological samples such as plasma, serum or tissues, unlike in SILAC which is applicable to cultured cells. The limitation in the technique lies in the back exchange of ^{18}O with naturally occurring ^{16}O (Back et al. 2002; Hicks et al. 2005). Thus, stable isotope labeling aims at both absolute and relative method of quantification, wherein the former relates to small-scale analyses of targeted specific proteins while the latter focused on large-scale global proteomic analyses. However, all techniques that rely on isotopic heavy/light labels for differential expression analysis compares two (or more) isotope envelopes. Under such circumstances, if a protein is completely repressed, it will result in the presence of only a single envelope which may be ignored depending on the sophistication of the analysis software. Ironically, the effect is silent to conditions that involve the transition from expression to complete repression.

6.5 Post-translational Modifications

One of the most challenging issues in proteomics with respect to human reproduction is the analysis of post-translational modifications (PTMs) of proteins. These
are in fact covalent processing events that have the capacity to modulate protein
properties by proteolytic cleavage or by the addition of a modifying group to one or
more amino acids. A variety of PTMs exists in nature with the most common being
phosphorylation, glycosylation, methylation, acetylation and acylation. These
modifications are known to regulate many physiological functions which involve
metabolic regulation via acetylation, regulation of enzyme activity in cellular signaling pathways via phosphorylation; cellular aging via oxidation; and the regulation of gene expression via methylation (Liu et al. 2013). Functional
characterization of these PTMs has revealed their role in a variety of cellular
processes including transcription, DNA damage, apoptosis, and cell-cycle regulation (Kouzarides 2007). Despite their great significance in various biological
functions, their study on a large scale has been disrupted by the lack of suitable
methods for its detection and thus only a few modifications have been discovered.
The major drawback lies in the fact that highly sensitive methods are required for its
identification because of its low stochiometry. For example, signaling kinase cascades are switched on and off by the reversible addition and removal of phosphorylated groups, as a result of which only 5–10 % of a protein kinase substrate is
phosphorylated, wherein methods are required to detect the modified protein at very
low levels (<5–10 fmol) (Mann and Jensen 2003; Seo and Lee 2004). Moreover,
determination of the exact position at which a typical protein is likely to undergo
post-translational modifications is an area of research that has posed a difficulty in
understanding using the conventional proteomic methods because of the following
reasons. Firstly, proteins undergo PTMs very frequently and its transient nature
requires prior information on its characteristics in order to identify a new PTM and
its position in the protein. Secondly, the labile nature of the covalent bond between
the PTM and amino acid side chain often poses a difficulty in maintaining the
peptide in its modified state during sample preparation and subsequent ionization in
MS (Seo and Lee 2004). However, of all the modifications described above,
post-translational regulation of proteins via protein phosphorylation is one of the
major means of protein regulation which has been widely studied. With enrichment,
phosphorylated PTMs can be detected with high sensitivity. Research on other
types of PTMs has been limited by the lack of enrichment methods because each
method can only identify one type of PTM (Zhao and Jensen 2009). Several
methods have been established to detect such changes, which include immunoprecipitation, chemical derivatization, affinity purification as well as the immobilized metal affinity chromatography (IMAC) methods. Analyzing PTMs is a
daunting task and only recently, the highly challenging task of revealing and
characterizing the dynamic protein phosphorylation networks has begun to become
feasible. Phosphoprotein analysis through mass spectra approach is further complicated by the technical difficulties of detecting phosphoproteins in the presence of

non-phosphorylated species. Immunoprecipitation is a commonly used tool in the analysis and identification of phosphoproteomics wherein proteins from complex mixtures are immunoprecipitated with antibodies against phosphorylated amino acid residues such as serine, tyrosine and threonine residues (Ignatoski 2001). This is followed by the separation by 1D-PAGE or 2D-PAGE gels and analyzed by MS to map the phosphorylation site. Although this method is advantageous in the identification and characterization of phosphorylated proteins and its residues, the methodology does not permit the quantification of phosphorylated proteins. Similarly, chemical derivatization is an approach which aids in the phosphoprotein enrichment through the β-elimination of phosphoric acid of pSer and pThr residues and the introduction of biotin moiety and the selective separation by chromatography from non-phosphorylated species. This too has its own limitations because of the loss of chromatographic performance and loss of sensitivity during mass spectral analysis and is thus restricted only to pSer and pThr residues (Goshe et al. 2001). Affinity purification is an alternative approach which involves metal oxides in the enrichment of phosphoproteins. Though this method is targeted to the enrichment of mono and multiply-phosphorylated proteins, and its recovery being 90 % for phosphopeptides, its application still needs to be thoroughly investigated for complex mixtures. Several selective enrichment techniques take advantage of the chelating properties of some metals toward the phosphate group of phosphorylated peptides, as in IMAC. IMAC too is an alternative approach which has been extensively used for the enrichment of phosphoproteins and peptides and targets mainly the multiply-phosphorylated proteins. This method involves the high affinity of the phospho-moiety to positively-charged metal ions like Fe^{3+}, Ga^{3+}, Al^{3+}, and Zr^{4+}, which are then immobilized on a solid support (silica, Sepharose, or agarose) with metal-chelating agents such as iminodiacetic acid (IDA), nitrilotriacetic acid (NTA), tris (carboxy methyl) ethylene diamine (TED), or poly (glycidyl methacrylate/divinyl benzene) (GMD) (Feng et al. 2007). Notably, the selectivity and specificity in IMAC can be altered with different buffer conditions, such as pH, salt concentration, buffer composition, and the presence of detergents (Jensen and Larsen 2007; Tsai et al. 2008). The major drawback in the method lies in the fact that the acidic groups on the peptides are non-specifically bound to the metal ions and hence need to be further esterified to block the acidic group, increasing the complexity of the experimental procedure. In addition to it, quantitative information is not obtained. PTMs of protein call the covalent modifications of amino acids collectively, resulting in the change of mass of the amino acid residue which is reflected by the change in the peptide's MS/MS spectrum. Thus, the characterization of PTMs is possible with MS/MS, but with a few difficulties.

Since PTMs are added after the translation from mRNA, identification of the modification sites is not easy, as a result of which protein databases usually do not contain information on the PTMs. PTM research has further been enhanced by the dozens of expert algorithms that have been created to perform unrestrictive searches, which can find almost all PTMs and even novel PTMs. Although computational analysis is advantageous in many ways, researchers often do not know which PTMs exist in their sample. Secondly, letting the software determine all the known

PTMs makes the computational analysis infeasible. Similarly, increasing PTMs search also significantly increases the false discoveries because of the growth of the searching space. A potential solution to this problem is to let the software identify the possible PTMs automatically from the data (Kumar and Mann 2009; Arnaudo and Garcia 2013).

6.6   Data Analysis

One of the major limitations of computational proteomics in complex samples is the sorting of information through the huge amount of generated data. Several algorithms and bioinformatics resources exist for the analysis of data resulting from proteomic analysis. Data analysis begins with spot identification in gel-based proteomics, which can be achieved with the help of available software followed by background correction and data normalization, before statistical analyses are performed. A typical proteomic analysis produces millions of spectra which can be analyzed via database searches that lead to the identification of a great number of false positives. Despite the availability of many software tools, separating actual identifications from false positives requires manual interpretation which is labor-intensive (Fu and Qian 2014). Data interpretation through SEQUEST further complicates the analysis because the resulting cross-correlation score (XCorr) is larger for larger peptides. The identification of hypothetical proteins are often incorrect which may be plausible because of the following reasons—poor quality of the spectrum, inaccuracy of spectrum prediction or imperfect scoring function as well as the degree of unknown PTMs. Nonetheless, with the development of recent software algorithms, interpretation of proteomic results have improved through the removal of low quality MS/MS spectra, post-search filtering of identifications and controlling the accepted error rate of identification through automated validation of protein identification. Whatever the case may be, the softwares developed thus far still require the integration of proteomics quantitation software and the direct accessibility of all proteomics results through a public database (Tabb et al. 2002; MacCoss et al. 2003).

Proteomics is very much a technology-driven field and its application in the field of reproductive biology using MS analysis is in the identification of protein or PTM biomarkers. In fact, the field of proteomics has observed a rapid growth over the last decade, driven strongly by improvements in MS analysis and the bioinformatics platform that supports it. This allows the analysis of complex reproductive samples from being labor-intensive to computational-intensive. Despite these technological advances and methodological breakthroughs, progress in the area of human reproduction has not dramatically taken off and there should to be putative identification that can serve as potential biomarkers. Challenges exist in almost all steps of proteomics identification. Biomolecules such as proteins are known to show a great deal of diversity, which has made their detection an ambitious feat. In order to achieve this goal, integration of different analytical platforms that enables a maximum range of analyte detection and identification through high sensitivity, selectivity and resolution is required. In proteomics, the sensitivity and dynamic range are limited by the efficiency of the LC–MS interfaces and the ion capacity of mass analyzers. To note, the quality of acquired MS/MS spectra is of great importance for successful identification, while its heterogeneity poses a major challenge for effective usage of spectral libraries. In this context, it is again note-worthy to mention that different technological advances with regards to proteomics identification sometimes seem to be in competition with each other. However, the key role lies in understanding the strengths and weaknesses of different technology and combining the different forces to understand complex protein structure and its modifications. Furthermore, as mentioned earlier, the key feature of proteins is its greater diversity. While until recently, the main focus of biomedical research has been genome-based, and most ontologies and annotations are gene-centric which often fails to capture protein-specific characteristics and function. Proteomics holds much promise in the field of reproductive biology. New types of proteomics technology and methodological breakthroughs, in combination with advanced bioinformatics, is currently on the rise to identify candidate biomarkers in the field of human reproduction based on protein pathways and signaling cascades. In this context, it is therefore envisaged that future development should focus to tune the available resources, in order to fulfill the needs of proteomic research in the field of human reproduction.

Chapter 7
What Does the Future Hold

Ricardo P. Bertolla

Proteomics is a rapidly evolving area within a global systems biology cascade, in which technological advances have greatly impacted data generation, mining, and interpretation up- and downstream, integrating knowledge from the genome down to metabolites and small molecules (Zhang et al. 2013; Weckwerth 2011; Rouillard et al. 2015; Marcoux and Cianferani 2015). Current proteomics studies are capable of generating a large amount of data for individual patients, and developments in filtering information out of this data is one of the main challenges for mass spectrometry-based studies (Zhang et al. 2013; Marcoux and Cianferani 2015; Bantscheff et al. 2012).

Moreover, data-intensive studies bring a promise of a holistic view of the biological system, as opposed to the conventional reductionist Cartesian view (Boogerd et al. 2007). This reductionism-holism debate, in which systems biology represents a general view of the system (holism), while molecular studies on specific pathways represent a reductionist view, is one of the most important methodological questions brought upon by advances in analytical and data processing technologies (Mazzocchi 2012; Merelli et al. 2014).

On the one hand, Cartesian reductionism may not be able to demonstrate the biological complexity responsible for subtle phenotypic differences, nor handle the difficulty of integrating environmental actions on this phenotype. On the other hand, a holistic view may derive from and generate biased information, as the vast amount of information required is prone to statistical demonstrations which are not *de facto* observable or repeatable (Mazzocchi 2012). It remains an important challenge therefore, to integrate these two seemingly paradoxical views in order to generate scientific information—one adds on to the other, so to say.

In addition, recent advances in targeted mass spectrometry-based proteomics, such as in selected reaction monitoring (SRM), multiple reaction monitoring (MRM), parallel reaction monitoring (PRM), and in data independent acquisition (DIA), bring

Ricardo P.Bertolla (✉)
São Paulo Federal University, São Paulo, São Paulo, Brazil
e-mail: rbertolla@yahoo.com

© The Author(s) 2016
A. Agarwal et al., *Proteomics in Human Reproduction*,
SpringerBriefs in Reproductive Biology, DOI 10.1007/978-3-319-48418-1_7

the promise of high-fidelity quantification of peptides/proteins in a biological sample (Lange et al. 2008; Gallien et al. 2015; Lo Turco et al. 2010; Camargo et al. 2013). Developments in both untargeted and targeted top-down proteomics (analysis of whole proteins in mass spectrometry) may simplify sample preparation and decrease inter-assay variation, as well as allow using imaging mass spectrometry to observe whole protein localization (Tran et al. 2011; Savaryn et al. 2013).

In this chapter, we will discuss how current studies in reproductive medicine have benefited from shotgun proteomics techniques, and how new developments are underway to deal with some of the shortcomings in such studies. We will also discuss how these findings have been applied to the field of reproductive biology, and possibilities for translation of these findings into a clinical setting. Finally, we will briefly discuss integration of proteomics within a global "Omics" cascade.

7.1 Benefits/Advantages of Proteomics in Reproductive Medicine

Shotgun proteomics allows the identification of a large number of proteins in any given sample (Zhang et al. 2013). Current liquid chromatography-tandem mass spectrometry (LC-MS/MS) techniques have increased dynamic range, which is especially important in samples in which a few proteins are highly enriched (Bantscheff et al. 2012). As a transudate of blood, the follicular fluid is an example of such a matrix. Albumin alone may account for over 50% of the follicular fluid proteome, and increased dynamic range has allowed identification and quantification of over 400 proteins, without depletion of enriched proteins (Lo Turco et al. 2010). Moreover, the follicular microenvironment is rich in exosomes and microvesicles, which transport, among other molecules, proteins to the oocyte and its surrounding cumulus cells, and shotgun proteomics is an interesting approach to identify and quantify contributing proteins to oocyte biology (da Silveira et al. 2012, 2015; Santonocito et al. 2014).

Seminal plasma is constituted of secreted fluids from the prostate, seminal vesicles, and the vas deferens/epididymal lumen/seminiferous tubule lumen (World Health Organization 2010). In seminal plasma, 15 proteins constitute almost 90% of the seminal plasma proteome (*unpublished results*), and the use of shotgun proteomics with a high dynamic range is also essential in order to allow for accurate identification and quantification of proteins (Camargo et al. 2013; Intasqui et al. 2013b; da Silva et al. 2013). Because the vast majority of the ejaculate volume in humans originates from the seminal vesicles and prostate (World Health Organization 2010), this is especially important in studies focusing on proteins that reflect testicular function, as these proteins will be diluted in the ejaculate, and testicular proteins may only account for 10% of the total seminal fluid proteome (Batruch et al. 2012). However, it is noteworthy that the epididymis presents a high concentration of exosomes (epididymosomes), which reflect final sperm maturation within the post-testicular environment (D'Amours et al. 2012).

Proteomics of spermatozoa is also an interesting area of research, as the search for membrane markers of sperm integrity or quality will allow for selection of sperm based on a functional trait, rather than relying on sperm motility or vitality (Vasen et al. 2015; Said et al. 2006; Intasqui et al. 2013a). Moreover, sperm proteomics is in itself a true reflection of testicular function and output, and may provide a deeper insight into testicular effects of diseases and/or their treatments (Intasqui et al. 2013a).

Studying the secreted proteins of embryos in spent culture media is another area of research that is interesting because it may, in the future, direct embryo selection based on its metabolism and proteomic pathway, rather than on morphology (Mains et al. 2011; Cortezzi et al. 2011; Krisher et al. 2015). An important technical limitation exists though, because the secreted proteins will be present at a very low concentration when compared to the supplemented albumin (HSA), and depletion of albumin or enriching proteins in lower amount may be fundamental for a comprehensive observation of the embryo secretome (Nyalwidhe et al. 2013). Moreover, a number of different proteins are present in unconditioned embryo culture media, and this should also be considered (Dyrlund et al. 2014).

Because reproductive biology is a complex system with several different biological states, in which fertility is defined as the interaction between a male and a female fertility potential, development of highly sensitive and specific technologies is of the essence. High sensitivity will allow for detection of alterations as soon as they are established, while high specificity will allow for differentiating phenotypic differences not easily assessable, such as in differentiating a high quality from a low quality, morphologically mature normal oocyte, or diagnosing an adolescent varicocele that will indeed lead to infertility in the future, as opposed to an innocuous varicocele.

7.2 Dealing with Limitations in the Use of Proteomics in the Field of Reproduction

Despite the accumulating number of studies with lists of proteomes from different cell types, systems, fluids and organisms, translation of these results into biomarkers or targets remains a difficult and costly process (Fuzery et al. 2013). Dealing with these limitations is of increasing importance in order to validate the large number of lists of differentially expressed proteins, as well as to apply the generated results into a practical clinical setting.

One important limitation in shotgun proteomics experiments is sample dynamic range and complexity. Sample dynamic range issues arise due to the presence of highly enriched proteins, such as albumin in follicular fluid (Lo Turco et al. 2010) or semenogelins I and II in seminal plasma (Camargo et al. 2013; Del Giudice et al. 2013; Zylbersztejn et al. 2013; Intasqui et al. 2013b). Sample complexity issues arise because gel-free digestion of a cell or fluid proteome leads to the formation of a large number of peptides (up to five orders of magnitude) (Michalski et al. 2011), which, if directly injected into mass spectrometers, leads to ionic suppression (some peptides will ionize more efficiently than others, thus suppressing the latter) and errors in data-dependent acquisition of the tandem mass spectra for peptide identification (Michalski et al. 2011).

Using modern ultra-high pressure liquid chromatography (UPLC) is very efficient in separating these peptides and decreasing sample complexity (peptide prefractionation) and, while co-elution of peptides may still occur, albeit at a much smaller rate, gel-free liquid chromatography—tandem mass spectrometry (LC-MS/MS) has generated much information in proteomics (Zhang et al. 2013; Camargo et al. 2013; Agarwal et al. 2016b; Intasqui et al. 2013a, b; Cox and Mann 2011). It remains important to increase dynamic ranges, however, because if only proteins higher in concentration are observed, sensibility in determining mild phenotypic alterations is greatly decreased, as differential expression in these cases are more often than not confined to proteins in lower amounts or of less network centrality, due to the high lethality of proteins high in quantity or centrality (Jeong et al. 2001).

Another approach to decrease sample complexity is to prefraction proteins, and then submit these generated fractions into a LC-MS/MS setup. In this type of experiment, proteins may be pre-fractionated based on their masses, using one-dimensional polyacrylamide gel electrophoresis (1D-PAGE), and slicing the protein lanes into sections for further in-gel digestion and injection into the LC-MS/MS system (Kim et al. 2003). Another option is to prefraction proteins based on their isoelectric point (the pH at which their net charge is zero), using either in-gel or in-solution isoelectric focusing (IEF) (Pernemalm and Lehtio 2013; Zuo and Speicher 2002).

A further advantage of this approach is that, in samples in which a single protein or group of proteins is highly enriched, as is the case with follicular fluid and seminal plasma for example, prefractionation allows isolation of these proteins to within a mass or pH range, further enriching proteins present in lower amounts in other ranges. In other words, protein prefractionation increases the dynamic range in which a proteome is able to be assessed. Limitations of protein prefractionation lie in the fact that increased processing is necessary, which in turn inserts variability to the study. This has been dealt with by utilizing labeling of whole proteins and pooling the study and control samples for direct comparison (Hoedt et al. 2014; Ong et al. 2002; Chahrour et al. 2015).

A second important limitation lies in confirmatory proteomics. Confirmation using Western blotting (probing of proteins separated by 1D-PAGE and transferred to membranes) is widely performed and, while not quite practical for experiments in which the number of differentially expressed proteins is high, is also widely accepted (Rifai et al. 2006; Surinova et al. 2011; Kohler and Seitz 2012). On the other hand, Western blotting is limited by cross-reactivity of the primary antibody and, quite often, by selection of candidates for confirmation (Alegria-Schaffer et al. 2009). Moreover, proteomes may be altered before they are accessible for study, which may impair conventional Western blot-based studies.

For example, when we consider seminal plasma in humans, semen liquefaction originates from proteolytic activity of kallikrein-III (KLK3). KLK3 is inactivated by Zn^{2+} after semen liquefaction, which in turn is avidly bound by the semen coagulation proteins semenogelins I and II. Upon binding of Zn^{2+} by semenogelins, KLK3 is activated, hydrolyzing the semenogelins and, thus, the seminal coagulum. Therefore, KLK3's proteolytic activity is mostly limited to semenogelins (Robert and Gagnon 1999; Mitra et al. 2010; Tomar et al. 2013). However, KLK3 does present a Trypsin-like cysteine/serine peptidase domain, which may lead to proteolysis of other proteins in the seminal plasma, thus altering protein mass in one-dimensional gels and, potentially, their response to antibody probing (UniProt).

Utilization of mass spectrometry-based targeted quantification of proteins, as discussed in Chaps. 2 and 6, may be an interesting approach in order to confirm results from shotgun proteomics experiments. While still a long pipeline, current developments in PRM and integrating results from quadrupole-orbitrap hybrid mass spectrometers (such as in the Thermo Q-Exactive mass spectrometer) may allow for shortening of the path and to construct PRM experiments from existing shotgun proteomics data (Law and Lim 2013; Gallien et al. 2015; Lange et al. 2008). The Skyline application developed by the MacCoss Lab Software offers an interesting pipeline for such a discovery-to-validation pipeline (MacLean et al. 2010).

A third limitation of proteomics experiments lies in the normalization of protein quantification. For distinct cell types, normalization to constitutive proteins is generally accepted as a valid approach, lending the concept from gene expression studies (Taylor and Posch 2014). Thus, normalizing protein quantification results to beta-tubulin or actin allows correcting for distortions in injection volume and minor pipetting errors. For follicular fluid however, it should be considered that much of what is important for follicular and oocyte maturation arises from the granulosa cell exudate that is secreted into the blood transudate that leads to antrum formation and growth as the follicle develops (Lo Turco et al. 2010; Rodgers and Irving-Rodgers 2010). Thus, it may be argued that normalizing to constitutive proteins in blood plasma may lead to distortions as to what is being secreted, as increases in volume (and thus in participation of blood plasma proteins) may dilute important proteins secreted into the follicular environment. Some groups have tackled this issue by studying proteins in microvesicles and exosomes present in the follicular fluid, as these have probably arisen from the surrounding cells (Santonocito et al. 2014; da Silveira et al. 2012, 2015).

Seminal plasma, on the other hand, may impose a greater challenge, due to peculiarities in its origin. Almost 75% of the ejaculate volume arises from the seminal vesicles, and nearly 25% from the prostate. Fluid from the seminiferous tubule lumen/epididymis/vas deferens contributes with very little volume to the final ejaculate (around 1–5%) (Plant and Zeleznik 2014), but is where proteins reflective of testicular function and sperm maturation will be observed (Batruch et al. 2012). Thus, normalization of protein quantification results to conventional constitutive proteins present in semen, may lead to very important distortions in the amount of proteins lower in concentration originating from the testes. Normalizing results to known constitutive testicular/epididymal proteins is a possibility, as it may be reflective of the true testicular output, but this remains an approximation that must be dealt with in the future.

Recently, development of LC-MS/MS studies with a top-down proteomics approach, in which proteins are analyzed without previous digestion (as opposed to the bottom-up analysis of peptides) has brought the possibility of analyzing individual protein molecules in any given samples (proteoforms) (Savaryn et al. 2013). This has been applied to a shotgun-type approach in order to identify a number of proteoforms, and may be an interesting platform for whole proteome interrogation, especially focused on detecting post-translational modifications and isoforms (Tran et al. 2011).

Continuous development of top-down approaches may allow further insights into cellular localization of whole proteins, such as in mass spectrometry imaging (MSI) of whole peptides and proteins (Andersson et al. 2008). MSI is a rapidly growing field within mass spectrometry, which allows localization of biomolecules within tissues. A number of different techniques are currently available for MSI (de Rond et al. 2015), but they incur the same problems when dealing with sample complexity, and there is an important trade-off between sensitivity and imaging resolution, and current resolution limits are down to 20 μm at best (5 μm for lipids and metabolites), well above the optical limit of $\sim$ 180 nm and larger than a sperm head (Aichler and Walch 2015). However, MSI for a targeted top-down detection of specific proteins is a promising area of development for rapid detection of specific proteins (and their post-translational modifications) in cells and tissues (de Rond et al. 2015; Aichler and Walch 2015).

7.3 Translating Findings into a Clinical Setting

Proteomics has the potential to impact many aspects of the clinical management of infertility, ranging from improved diagnosis, establishment of prognosis, and in offering a personalized treatment for the infertile couple. Individualization of disease is an increasing topic of interest as a more personalized medicine is evolving, and proteomics offers a possibility of differentiating subtle phenotypic differences in establishing a diagnosis (Merelli et al. 2014; Hood and Flores 2012; Williams and Auwerx 2015; Yan 2014). For example, varicocele is a disease which is present in 15% of adult men, 35% of men with primary infertility, and in 80% of men with secondary infertility (Gorelick and Goldstein 1993; Cozzolino and Lipshultz 2001). However, 80% of men with varicocele will not present with infertility, which demonstrates the need to distinguish the presence of a varicocele that is indeed leading to altered spermatogenesis (Choi and Kim 2013). While many efforts have been made toward identifying cellular (Blumer et al. 2008, 2012; Lacerda et al. 2011; Padron et al. 1997; Fariello et al. 2012) and molecular markers of altered spermatogenesis (Camargo et al. 2013; Del Giudice et al. 2010; Fariello et al. 2012; Agarwal et al. 2015b, d, 2016a), it may be the case that these markers are identified only after irreversible alterations have been established within the seminiferous tubules. Identification of early markers of altered spermatogenesis, on the other hand, may allow for intervention before these alterations settle in (Del Giudice et al. 2013; Zylbersztejn et al. 2013).

Differentiating a high quality embryo from a low quality one is another area where proteomics analysis may play a role. While some technological advances have been made in terms of determining aneuploidy and even some genetic mutations associated to congenital abnormalities (Brezina and Kutteh 2015; Gardner et al. 2015), studying the secreted proteins from an early embryo may indeed reflect its metabolic state, leading to improved embryo selection for couples undergoing assisted reproduction (Cortezzi et al. 2011; Krisher et al. 2015; Benkhalifa et al. 2015; Nyalwidhe et al. 2013). Coupled with proteomics analysis of the oocyte secretome and of its surrounding cumulus cells and follicular fluid, these technologies may indeed shift the paradigm in terms of assessing early embryo biology and quality (Montag et al. 2013). Moreover, understanding the secretome of healthy oocytes (and their microenvironment) brings the promise of developing improved media for in vitro follicular and oocyte maturation.

Selection of higher quality gametes will also gain from current developments in proteomics technologies. Determining sperm surface markers of low quality, such as sperm DNA fragmentation for example, may allow for removal of affected sperm from a sample without resorting to centrifugation, as is currently performed, which is an important source of iatrogenic oxidative damage to sperm (Zini et al. 1999, 2000; Stevanato et al. 2008). These protein markers may be bound by labeled antibodies and sorted utilizing fluorescence-activated cell sorting (FACS) or magnetic-activated cell sorting (MACS). This has been tested for in the past using annexin-V, a marker of initial apoptosis, and has rendered positive results (Said et al. 2006). Indeed, a panel of selective surface biomarkers may be developed, in order to sort out sperm with alterations to a number of functions, such as DNA fragmentation, mitochondrial alterations or acrosome damage.

Finally, it should be mentioned that the offer of a personalized medicine is increasingly important in order to determine optimal treatment options for infertile couples (Hood and Flores 2012). The use of proteomics (and other analyses in the Omics cascade) may indeed allow for determination of improved hormonal stimulation regimes for infertile women, a targeted approach for improved spermatogenesis in men and improved selection and culture conditions for early embryos.

7.4 Systems Biology and the Omics Cascade

As the concept of personalized medicine is still evolving, developments in understanding the full Omics cascade, ranging from genomics (gene mutations, epigenetics, imprinting) and transcriptomics through proteomics down to metabolomics (Hood and Flores 2012; Ferreira et al. 2010), have greatly increased the generated data from each individual patient. Integrating multiscalar data into comparable datasets is one of the challenges that this approach brings, but the increasing capability of modern equipment in detecting, with high resolution, small molecules and peptides associated to next-generation sequencing allows for a full map of the biological networks active in any given biological condition (Merelli et al. 2014; Rouillard et al. 2015).

Under this approach, gene stability/silencing/activation/polymorphism information is overlaid to complete transcriptomes, generating gene expression profiles, as well as RNA-level control of translation. These are further understood when quantitative shotgun proteomics studies demonstrate turnover of these gene products into proteins which, integrated with genomics information, may contain specific amino acid alterations. If these are present in specific domains, protein function may potentially be altered, as may protein conformation. These proteins are further understood as protein–protein interaction networks that participate in specific pathways, sharing functions and domain activity in order to affect these biological pathways. Moreover, protein function is finely tuned by post-translational modifications, such as phosphorylation in sperm capacitation. Downstream metabolites are substrates, but are also generated by protein activity, and regulate upstream gene expression (through epigenetic imprinting), translation and protein function (Hood and Flores 2012; Ferreira et al. 2010; Rouillard et al. 2015; Lausted et al. 2014). Phosphorylcholines, for example, are known to interact with specific epididymal proteins (released into the epididymal fluid) and activate their docking onto the sperm phospholipid membrane (D'Amours et al. 2012).

The study of metabolites, metabolomics and of different post-genomic molecules, such as lipids (lipidomics), sugars (glycomics) and steroid hormones (steroidomics), among others, is a rapidly growing area of research which has greatly benefited from advances in mass spectrometry (Ferreira et al. 2010; Patel and Ahmed 2015), while development of miniaturization technologies may benefit the clinical practice with the so-called "Lab-on-a-chip" solution for rapid mass spectrometry-based clinical diagnosis (Oedit et al. 2015). There are currently specific software platforms designed to integrate these data into biological networks, which may shed some light on current understanding of disease development as well as on homeostasis of biological systems (Cline et al. 2007; Xia et al. 2012).

In the reproductive system, this is particularly interesting because of the myriad of biomolecules affecting signaling pathways and modulating the full range of genomics-to-metabolomics response at play in any given biological condition. This approach has been applied to understanding endometrial receptivity, for example (Minten et al. 2013; Spencer et al. 2013). Mass spectrometry-based lipid profiling has also been applied to follicular fluid (Cataldi et al. 2013; De Oliveira et al. 2012), semen (Camargo et al. 2014; Glander et al. 2002; Schiller et al. 2000), and even animal embryos under different culture conditions (Ferreira et al. 2010), demonstrating the ability to differentiate diverse conditions in both cases.

The biggest challenge these studies bring is filtering out background noise from true biological signals when integrating these multiscalar data (Xia et al. 2012; Rouillard et al. 2015; Cline et al. 2007; Merelli et al. 2014). Much has been developed in terms of applying statistical learning tests. Moreover, multivariate data analysis utilizing Principal Component Analysis (PCA) and Partial Least Squares Discriminant Analysis (PLS-DA), originally derived from chemometrics studies, have been shown to be useful approaches to simplify data analysis (Xia et al. 2012; Camargo et al. 2014; Intasqui et al. 2015). Coupled with tests developed for microarray studies, such as Significant Analysis of Microarrays (SAM) and Empirical Bayes Analysis of Microarrays (EBAM), and discriminant analyses, such as logistic and multinomial regression, and cluster analysis, as well as with conventional univariate tests, data filtering and mining has been able to generate biological targets for downstream validation as disease biomarkers.

7.5 Final Remarks

The future of proteomics in the field of reproductive biology will be likely characterized by a rapid increase in sensitivity and resolution of analytical techniques (Wang et al. 2014a; Junger and Aebersold 2014; Zhao et al. 2014; Fuhrer and Zamboni 2015), as well as in a decrease in equipment size and cost (Gao et al. 2006; Blain et al. 2004), a parallel to Moore's Law in mass spectrometry. Computational solutions for complex data analysis will also play an important role in translating these findings into actionable sources of information for a practical impact (Cox and Mann 2008; Cline et al. 2007; Xia et al. 2012; Haga and Wu 2014; Maere et al. 2005; Mi et al. 2013). Moreover, the potential development of ambient ionization sources for mass spectrometers coupled with improvement in targeted top-down proteomics may indeed allow screening for protein biomarkers in the clinic or during a surgical procedure (Hsu and Dorrestein 2015). A similar approach has been utilized, for example, for mapping of a metabolite marker of cancer during brain tumor surgery (Santagata et al. 2014).

Furthermore, disease individualization will definitely benefit greatly from the development of proteomics-based biomarkers (Hood and Flores 2012), with a direct impact in the clinical management of the infertile couple as well as on sperm and oocyte selection and maturation, and in early embryo culture and selection. Determination of potential supplements for gamete and embryo cryopreservation or for other processing technologies is also promising if current shotgun studies map out determinants of cell integrity in these conditions. However, overcoming the hurdle of big data analysis is of the essence, so as to not find ourselves swarmed with data but with very few information to act upon (Merelli et al. 2014).

In conclusion, in the search for the true biology of a given system when undertaking studies within the systems biology multi-platform setting, one must not wander too far from the essence of the science pursued. Biology, in truth, does not follow holism, nor does it follow Cartesian reductionism—these are rather interpretations that allow comprehension of the presented fact (Boogerd et al. 2007). In this sense, while a holistic view of the biological system is a promise of current Omics-based platforms, reductionism remains the driving force for research and development in reproductive biology, and a conventional workflow based on hypothesis generation following a Cartesian approach and data validation will remain of essence in knowledge generation—even if the data are generated in a shotgun-type study. This is especially true in large datasets, in which a flawed design or a biased approach will likely find one or another statistical validation but will not report the underlying truth. The most important benefit that shotgun proteomics—and Omics studies in general will be to thoroughly answer well-designed questions and hypotheses.

Chapter 8
Conclusions

Luna Samanta and Damayanthi Durairajanayagam

Proteomics, the high-throughput measurement technologies, in which structure or function of proteins are studied on a global scale, are opening wider doors into reproductive medicine and technology. In the post-genomic era, studies have mainly focused on the identification of novel protein biomarkers in complex biological systems. A biomarker is a distinctive biological or biologically-derived indicator of a process, event, or condition, and is considered ideal if it serves the purpose of screening, diagnosis, and monitoring disease activity. In addition, they may be held responsible for targeted therapy or could even assess therapeutic responses. With regard to infertility, the main objective of a biomarker is to evaluate in an accurate and minimally invasive manner, the potential of a couple to conceive a child.

This book aims to be an innovative discourse on proteomics and its application in human reproduction. Beginning with the basics, the methods used in proteomics studies, i.e. the typical flow in a proteomics experiment, from sample preparation to separation techniques, protein identification via mass spectrometry, data analyses, and functional enrichment and validation tests was described in the Introduction chapter. The following three chapters each presented a comprehensive perspective on proteomics studies on the three main concerns of human reproduction respectively, i.e., male infertility, female infertility, and assisted reproduction. Clearly, the employment of proteomics methodologies provides an opportunity to examine infertility at a molecular level.

L. Samanta (✉)
Department of Zoology, Ravenshaw University, Cuttack, India
e-mail: lsamanta@ravenshawuniversity.ac.in

D. Durairajanayagam
Faculty of Medicine, Universiti Teknologi MARA, Sungai Buloh Campus,
Selangor, Malaysia
e-mail: damayanthi.d@gmail.com

© The Author(s) 2016
A. Agarwal et al., *Proteomics in Human Reproduction*,
SpringerBriefs in Reproductive Biology, DOI 10.1007/978-3-319-48418-1_8

For the chapter on male infertility, proteins related to various infertility conditions that have been identified in the spermatozoa and in the seminal plasma were described. Using the data on differentially expressed proteins, proteins that are absent or expressed abnormally, could present a window into the etiology of infertility present. Moreover, comparative proteomic profiling of samples from fertile donors vs. that of infertile patients, anomalies in protein expression pertaining to a particular type of infertility, such as azoospermia, varicocele or asthenozoospermia, or even unexplained infertility may be detected. Similarly in the female infertility chapter, proteins associated with fertility in the reproductive microenvironments of females, such as endometrial fluid, peritoneal fluid, and follicular fluid, reproductive tissues and serum that could contribute valuable information toward the diagnosis of female factor disorders, such as endometriosis, PCOS, and unexplained infertility were described.

With regard to assisted reproduction, it is likely that information obtained using proteomics will revolutionize the way current IVF procedures are performed. This is because these Omics technologies are suitable diagnostic tools to explore differences among follicles, human gametes, and embryos. Moreover, since single embryo transfer is gaining momentum across the IVF clinics, application of such platforms will become inevitable in the selection of the embryos for transfer. In the chapter on proteomics in assisted reproduction, novel molecular biomarkers related to infertility problems have been described, allowing the expansion of knowledge in order to design new diagnostic or selection tests with the aim of improving the success rates of ART. By working out the involvement and contribution of the proteins of interest in gamete or embryo physiology, and subsequently monitoring the protein profile in infertile patients, proteins of interest could act as diagnostic biomarkers and perhaps even as therapeutic targets in the management of infertility.

In the antepenultimate chapter, the existing challenges encountered during the conduct of proteomic studies in reproductive medicine were described. These include technological limitations that arise due to the standard use of gel-based protein separation and protein identification by mass spectrometry that may end up discounting the presence of low abundance proteins. Advances in gel-free high-throughput technologies offer a possible solution to this issue. Another pertinent challenge faced in proteomics studies within the reproductive field is the innately heterogeneous nature of the samples studied. This could be contributed to the diverse structural and physiochemical characteristics of the proteins present. Ultimately, the vast dynamic range of these samples favors the identification of high abundance proteins at the expense of low abundance proteins. On top of the complexities of the sample, the individual variation within each sample contributes further to its intricacy. Besides this, the detailed technical aspect of a mass spectrometry analysis workflow, for example, does influence protein quantitation and thus the outcome of the study, which also contribute toward the low reproducibility of the protein list generated. Another crucial challenge faced is a shortage of highly sensitive methods for the identification of post-translational modifications of proteins. However of late, investigations that aim to characterize dynamic protein phosphorylation networks are underway. Lastly, the key limitation in applying computational proteomics to complex samples, which is to organize the information based on the enormous amounts of data generated, was highlighted.

The penultimate chapter of this book brought to light the positive outcomes of shotgun proteomics on studies in reproductive medicine, such as the ability to identify a large number of proteins. However, with the generation of extensive lists of proteins comes the challenge of interpreting these results in order to elucidate target proteins or biomarkers of interest as well as validating the relevant proteins that are differentially expressed with the goal of eventually translating these findings from a bench to a clinical setting. Limitations of shotgun proteomics studies, such as sample dynamic range and complexity issues, and possible problems arising from the use of Western blotting for confirming the proteins identified, and the practice of normalizing protein quantification results to constitutive proteins were also highlighted. Lastly, the integration of these proteomics findings into the global Omics cascade was also touched upon.

Exploration of the molecular mechanisms underlying infertility in males and females using mass spectrometry-based quantitative proteomics and related Omics technologies has a crucial role in the future of reproductive research. In future, these studies may provide the evidence needed to finally decipher the pathophysiology and molecular pathways leading up to infertility.

References

Abrao MS, Muzii L, Marana R (2013) Anatomical causes of female infertility and their management. Int J Gynaecol Obstet: Official Organ Int Fed Gynaecol Obstet 123(Suppl 2): S18–S24. doi:10.1016/j.ijgo.2013.09.008

Aebersold R (2003) A mass spectrometric journey into protein and proteome research. J Am Soc Mass Spectrom 14(7):685–695. doi:10.1016/S1044-0305(03)00289-7

Aebersold R, Cravatt BF (2002) Proteomics–advances, applications and the challenges that remain. Trends Biotechnol 20(12 Suppl):S1–2

Aebersold R, Mann M (2003) Mass spectrometry-based proteomics. Nature 422(6928):198–207. doi:10.1038/nature01511

Agarwal A, Ayaz A, Samanta L, Sharma R, Assidi M, Abuzenadah AM, Sabanegh E (2015a) Comparative proteomic network signatures in seminal plasma of infertile men as a function of reactive oxygen species. Clinical proteomics 12 (1):23. doi:10.1186/s12014-015-9094-5

Agarwal A, Gupta S, Sharma RK (2005b) Role of oxidative stress in female reproduction. Reproductive biology and endocrinology: RB&E 3:28. doi:10.1186/1477-7827-3-28

Aggarwal K, Choe LH, Lee KH (2006) Shotgun proteomics using the iTRAQ isobaric tags. Briefings Funct Genomics Proteomics 5(2):112–120. doi:10.1093/bfgp/ell018

Agarwal A, Hamada A, Esteves SC (2012) Insight into oxidative stress in varicocele-associated male infertility: part 1. Nat Rev Urol 9(12):678–690. doi:10.1038/nrurol.2012.197

Agarwal A, Durairajanayagam D, Halabi J, Peng J, Vazquez-Levin M (2014) Proteomics, oxidative stress and male infertility. Reprod Biomed Online 29(1):32–58. doi:10.1016/j.rbmo. 2014.02.013

Agarwal A, Sharma R, Durairajanayagam D, Ayaz A, Cui Z, Willard B, Gopalan B, Sabanegh E (2015a) Major protein alterations in spermatozoa from infertile men with unilateral varicocele. Reprod Biol Endocrinol 13:8. doi:10.1186/s12958-015-0007-2

Agarwal A, Sharma R, Durairajanayagam D, Cui Z, Ayaz A, Gupta S, Willard B, Gopalan B, Sabanegh E (2015b) Differential proteomic profiling of spermatozoal proteins of infertile men with unilateral or bilateral varicocele. Urology 85 (3):580–588. doi:10.1016/j.urology.2014.11. 030

Agarwal A, Sharma R, Durairajanayagam D, Cui Z, Ayaz A, Gupta S, Willard B, Gopalan B, Sabanegh E (2015c) Spermatozoa protein alterations in infertile men with bilateral varicocele. Asian J Androl

Agarwal A, Bertolla RP, Samanta L (2016a) Sperm proteomics: potential impact on male infertility treatment. Expert Rev Proteomics 13(3):285–296. doi:10.1586/14789450.2016. 1151357

Agarwal A, Sharma R, Durairajanayagam D, Cui Z, Ayaz A, Gupta S, Willard B, Gopalan B, Sabanegh E (2016b) Spermatozoa protein alterations in infertile men with bilateral varicocele. Asian J Androl 18(1):43–53. doi:10.4103/1008-682x.153848

© The Author(s) 2016
A. Agarwal et al., *Proteomics in Human Reproduction*,
SpringerBriefs in Reproductive Biology, DOI 10.1007/978-3-319-48418-1

Agarwal A, Sharma R, Samanta L, Durairajanayagam D, Sabanegh E (2016c) Proteomic signatures of infertile men with clinical varicocele and their validation studies reveal mitochondrial dysfunction leading to infertility. Asian J Androl 18(2):282–291. doi:10.4103/1008-682X.170445

Ahmed FE (2009) Sample preparation and fractionation for proteome analysis and cancer biomarker discovery by mass spectrometry. J Sep Sci 32(5–6):771–798. doi:10.1002/jssc.200800622

Aichler M, Walch A (2015) MALDI Imaging mass spectrometry: current frontiers and perspectives in pathology research and practice. Lab Inv J Tech Methods Pathol 95(4):422–431. doi:10.1038/labinvest.2014.156

Aitken RJ, Nixon B, Lin M, Koppers AJ, Lee YH, Baker MA (2007) Proteomic changes in mammalian spermatozoa during epididymal maturation. Asian J Androl 9(4):554–564. doi:10.1111/j.1745-7262.2007.00280.x

Alegria-Schaffer A, Lodge A, Vattem K (2009) Performing and optimizing Western blots with an emphasis on chemiluminescent detection. Methods Enzymol 463:573–599. doi:10.1016/s0076-6879(09)63033-0

Alvarez Sedo C, Rawe VY, Chemes HE (2012) Acrosomal biogenesis in human globozoospermia: immunocytochemical, ultrastructural and proteomic studies. Hum Reprod 27(7):1912–1921. doi:10.1093/humrep/des126

Amanchy R, Kalume DE, Pandey A (2005) Stable isotope labeling with amino acids in cell culture (SILAC) for studying dynamics of protein abundance and posttranslational modifications. Sci STKE: Signal Transduct Knowl Environ 267:pl2. doi:10.1126/stke.2672005pl2

Amann RP (1989) Can the fertility potential of a seminal sample be predicted accurately? J Androl 10(2):89–98

Amaral A, Castillo J, Estanyol JM, Ballesca JL, Ramalho-Santos J, Oliva R (2013) Human sperm tail proteome suggests new endogenous metabolic pathways. Mol Cell Proteomics: MCP 12(2):330–342. doi:10.1074/mcp.M112.020552

Amaral A, Castillo J, Ramalho-Santos J, Oliva R (2014a) The combined human sperm proteome: cellular pathways and implications for basic and clinical science. Hum Reprod Update 20(1):40–62. doi:10.1093/humupd/dmt046

Amaral A, Paiva C, Attardo Parrinello C, Estanyol JM, Ballesca JL, Ramalho-Santos J, Oliva R (2014b) Identification of proteins involved in human sperm motility using high-throughput differential proteomics. J Proteome Res 13(12):5670-5684. doi:10.1021/pr500652y

Ambekar AS, Nirujogi RS, Srikanth SM, Chavan S, Kelkar DS, Hinduja I, Zaveri K, Prasad TS, Harsha HC, Pandey A, Mukherjee S (2013) Proteomic analysis of human follicular fluid: a new perspective towards understanding folliculogenesis. J Proteomics 87:68–77. doi:10.1016/j.jprot.2013.05.017

Ambekar AS, Kelkar DS, Pinto SM, Sharma R, Hinduja I, Zaveri K, Pandey A, Prasad TS, Gowda H, Mukherjee S (2015) Proteomics of follicular fluid from women with polycystic ovary syndrome suggests molecular defects in follicular development. J Clin Endocrinol Metab 100(2):744–753. doi:10.1210/jc.2014-2086

Amoako AA, Balen AH (2015) Female infertility: diagnosis and management. Endocrinology and Diabetes. Springer, pp 123–131

Andersen AN, Goossens V, Ferraretti AP, Bhattacharya S, Felberbaum R, de Mouzon J, Nygren KG, European IVFmC, European Society of Human R, Embryology (2008) Assisted reproductive technology in Europe, 2004: results generated from European registers by ESHRE. Hum Reprod 23(4):756–771. doi:10.1093/humrep/den014

Anderson NL (2010) The clinical plasma proteome: a survey of clinical assays for proteins in plasma and serum. Clin Chem 56(2):177–185. doi:10.1373/clinchem.2009.126706

Anderson NL, Anderson NG (2002) The human plasma proteome: history, character, and diagnostic prospects. Mol Cell Proteomics MCP 1(11):845–867

Andersson M, Groseclose MR, Deutch AY, Caprioli RM (2008) Imaging mass spectrometry of proteins and peptides: 3D volume reconstruction. Nat Methods 5(1):101–108. doi:10.1038/nmeth1145

Angelucci S, Ciavardelli D, Di Giuseppe F, Eleuterio E, Sulpizio M, Tiboni GM, Giampietro F, Palumbo P, Di Ilio C (2006) Proteome analysis of human follicular fluid. Biochimica et biophysica acta 1764(11):1775–1785. doi:10.1016/j.bbapap.2006.09.001

Antoniassi MP, Intasqui P, Camargo M, Zylbersztejn DS, Carvalho VM, Cardozo KH, Bertolla RP (2016) Analysis of the sperm functional aspects and seminal plasma proteomic profile from male smokers. BJU Int. doi:10.1111/bju.13539

Arnaudo AM, Garcia BA (2013) Proteomic characterization of novel histone post-translational modifications. Epigenetics Chromatin 6(1):24. doi:10.1186/1756-8935-6-24

Ashburner M, Ball CA, Blake JA, Botstein D, Butler H, Cherry JM, Davis AP, Dolinski K, Dwight SS, Eppig JT, Harris MA, Hill DP, Issel-Tarver L, Kasarskis A, Lewis S, Matese JC, Richardson JE, Ringwald M, Rubin GM, Sherlock G (2000) Gene ontology: tool for the unification of biology. Gene Ontology Consortium Nat Genet 25(1):25–29. doi:10.1038/75556

Atiomo W, Khalid S, Parameshweran S, Houda M, Layfield R (2009) Proteomic biomarkers for the diagnosis and risk stratification of polycystic ovary syndrome: a systematic review. BJOG Int J Obstet Gynaecol 116(2):137–143. doi:10.1111/j.1471-0528.2008.02041.x

Auger J, Kunstmann JM, Czyglik F, Jouannet P (1995) Decline in semen quality among fertile men in Paris during the past 20 years. N Engl J Med 332(5):281–285. doi:10.1056/NEJM199502023320501

Ayaz A, Agarwal A, Sharma R, Arafa M, Elbardisi H, Cui Z (2015) Impact of precise modulation of reactive oxygen species levels on spermatozoa proteins in infertile men. Clin Proteomics 12(1):4. doi:10.1186/1559-0275-12-4

Aziz N (2013) The importance of semen analysis in the context of azoospermia. Clinics 68(Suppl 1):35–38

Back JW, Notenboom V, de Koning LJ, Muijsers AO, Sixma TK, de Koster CG, de Jong L (2002) Identification of cross-linked peptides for protein interaction studies using mass spectrometry and 18O labeling. Anal Chem 74(17):4417–4422

Baggerman G, Vierstraete E, De Loof A, Schoofs L (2005) Gel-based versus gel-free proteomics: a review. Comb Chem High Throughput Screen 8(8):669–677

Bai J, Fu SH, Cai LL, Sun L, Cong YL (2009) [Identification of proteins in the seminal plasma of healthy fertile men by shotgun proteomic strategy]. Zhonghua nan ke xue =. Nat J Androl 15(4):297–309

Bai J, Fu SH, Cai LL, Sun L, Cong YL (2010) [Identification of differential proteins in the seminal plasma of healthy fertile and non-obstructive azoospermia men by shotgun proteomic strategy]. Zhonghua nan ke xue =. Nat J Androl 16(10):887–896

Baker MA, Naumovski N, Hetherington L, Weinberg A, Velkov T, Aitken RJ (2013) Head and flagella subcompartmental proteomic analysis of human spermatozoa. Proteomics 13(1):61–74. doi:10.1002/pmic.201200350

Balakier H, Xiao R, Zhao J, Zaver S, Dziak E, Szczepanska K, Opas M, Yie S, Librach C (2013) Expression of survivin in human oocytes and preimplantation embryos. Fertil Steril 99(2):518–525. doi:10.1016/j.fertnstert.2012.09.020

Bantscheff M, Schirle M, Sweetman G, Rick J, Kuster B (2007) Quantitative mass spectrometry in proteomics: a critical review. Anal Bioanal Chem 389(4):1017–1031. doi:10.1007/s00216-007-1486-6

Bantscheff M, Lemeer S, Savitski MM, Kuster B (2012) Quantitative mass spectrometry in proteomics: critical review update from 2007 to the present. Anal Bioanal Chem 404(4):939–965. doi:10.1007/s00216-012-6203-4

Barazani Y, Katz BF, Nagler HM, Stember DS (2014) Lifestyle, environment, and male reproductive health. Urologic Clin North Am 41(1):55–66. doi:10.1016/j.ucl.2013.08.017

Batruch I, Lecker I, Kagedan D, Smith CR, Mullen BJ, Grober E, Lo KC, Diamandis EP, Jarvi KA (2011) Proteomic analysis of seminal plasma from normal volunteers and post-vasectomy patients identifies over 2000 proteins and candidate biomarkers of the urogenital system. J Proteome Res 10(3):941–953. doi:10.1021/pr100745u

Batruch I, Smith CR, Mullen BJ, Grober E, Lo KC, Diamandis EP, Jarvi KA (2012) Analysis of seminal plasma from patients with non-obstructive azoospermia and identification of candidate biomarkers of male infertility. J Proteome Res 11(3):1503–1511. doi:10.1021/pr200812p

Bayasula Iwase A, Kobayashi H, Goto M, Nakahara T, Nakamura T, Kondo M, Nagatomo Y, Kotani T, Kikkawa F (2013) A proteomic analysis of human follicular fluid: comparison between fertilized oocytes and non-fertilized oocytes in the same patient. J Assist Reprod Genet 30(9):1231–1238. doi:10.1007/s10815-013-0004-3

Beall S, Brenner C, Segars J (2010) Oocyte maturation failure: a syndrome of bad eggs. Fertil Steril 94(7):2507–2513. doi:10.1016/j.fertnstert.2010.02.037

Beck M, Schmidt A, Malmstroem J, Claassen M, Ori A, Szymborska A, Herzog F, Rinner O, Ellenberg J, Aebersold R (2011) The quantitative proteome of a human cell line. Mol Syst Biol 7:549. doi:10.1038/msb.2011.82

Benkhalifa M, Madkour A, Louanjli N, Bouamoud N, Saadani B, Kaarouch I, Chahine H, Sefrioui O, Merviel P, Copin H (2015) From global proteome profiling to single targeted molecules of follicular fluid and oocyte: contribution to embryo development and IVF outcome. Expert Rev Proteomics 12(4):407–423. doi:10.1586/14789450.2015.1056782

Bianchi L, Gagliardi A, Campanella G, Landi C, Capaldo A, Carleo A, Armini A, De Leo V, Piomboni P, Focarelli R, Bini L (2013) A methodological and functional proteomic approach of human follicular fluid en route for oocyte quality evaluation. J Proteomics 90:61–76. doi:10.1016/j.jprot.2013.02.025

Bindea G, Mlecnik B, Hackl H, Charoentong P, Tosolini M, Kirilovsky A, Fridman WH, Pages F, Trajanoski Z, Galon J (2009) ClueGO: a Cytoscape plug-in to decipher functionally grouped gene ontology and pathway annotation networks. Bioinform (Oxford, England) 25(8):1091–1093. doi:10.1093/bioinformatics/btp101

Biomarkers Definitions Working G (2001) Biomarkers and surrogate endpoints: preferred definitions and conceptual framework. Clin Pharmacol Ther 69(3):89–95. doi:10.1067/mcp.2001.113989

Blain MG, Riter LS, Cruz D, Austin DE, Wu G, Plass WR, Cooks RG (2004) Towards the hand-held mass spectrometer: design considerations, simulation, and fabrication of micrometer-scaled cylindrical ion traps. Int J Mass Spectrom 236(1):91–104

Blumer CG, Fariello RM, Restelli AE, Spaine DM, Bertolla RP, Cedenho AP (2008) Sperm nuclear DNA fragmentation and mitochondrial activity in men with varicocele. Fertil Steril 90(5):1716–1722. doi:10.1016/j.fertnstert.2007.09.007

Blumer CG, Restelli AE, Giudice PT, Soler TB, Fraietta R, Nichi M, Bertolla RP, Cedenho AP (2012) Effect of varicocele on sperm function and semen oxidative stress. BJU Int 109(2):259–265. doi:10.1111/j.1464-410X.2011.10240.x

Boersema PJ, Raijmakers R, Lemeer S, Mohammed S, Heck AJ (2009) Multiplex peptide stable isotope dimethyl labeling for quantitative proteomics. Nat Protoc 4(4):484–494. doi:10.1038/nprot.2009.21

Boivin J, Bunting L, Collins JA, Nygren KG (2007) International estimates of infertility prevalence and treatment-seeking: potential need and demand for infertility medical care. Hum Reprod 22(6):1506–1512. doi:10.1093/humrep/dem046

Boogerd F, Bruggeman FJ, Hofmeyr J-HS, Westerhoff HV (2007) Systems biology: philosophical foundations. Elsevier

Bretveld R, Brouwers M, Ebisch I, Roeleveld N (2007) Influence of pesticides on male fertility. Scand J Work Environ Health 33(1):13–28

Brezina PR, Kutteh WH (2015) Clinical applications of preimplantation genetic testing. BMJ (Clin Res ed) 350:g7611. doi:10.1136/bmj.g7611

Brosens JJ, Hodgetts A, Feroze-Zaidi F, Sherwin JR, Fusi L, Salker MS, Higham J, Rose GL, Kajihara T, Young SL, Lessey BA, Henriet P, Langford PR, Fazleabas AT (2010) Proteomic analysis of endometrium from fertile and infertile patients suggests a role for apolipoprotein A-I in embryo implantation failure and endometriosis. Mol Hum Reprod 16(4):273–285. doi:10.1093/molehr/gap108

Brown JK, Lauer KB, Ironmonger EL, Inglis NF, Bourne TH, Critchley HO, Horne AW (2013) Shotgun proteomics identifies serum fibronectin as a candidate diagnostic biomarker for inclusion in future multiplex tests for ectopic pregnancy. PloS One 8(6):e66974. doi:10.1371/journal.pone.0066974

Bruce C, Stone K, Gulcicek E, Williams K (2013) Proteomics and the analysis of proteomic data: 2013 overview of current protein-profiling technologies. Curr Protoc Bioinform Chapter 13: Unit 13 21. doi:10.1002/0471250953.bi1321s41

Brucker C, Lipford GB (1995) The human sperm acrosome reaction: physiology and regulatory mechanisms.An update. Hum Reprod Update 1(1):51–62

Bungum M (2012) Sperm DNA integrity assessment: a new tool in diagnosis and treatment of fertility. Obstet Gynecol Int 531042. doi:10.1155/2012/531042

Byrjalsen I, Larsen PM, Fey SJ, Christiansen C (1995) Human endometrial proteins with cyclic changes in the expression during the normal menstrual cycle: characterization by protein sequence analysis. Hum Reprod 10(10):2760–2766

Cadavid JA, Alvarez A, Markert UR, Cardona Maya W (2014) Differential protein expression in seminal plasma from fertile and infertile males. J Hum Reprod Sci 7(3):206–211. doi:10.4103/0974-1208.142485

Cakmak H, Taylor HS (2011) Implantation failure: molecular mechanisms and clinical treatment. Hum Reprod Update 17(2):242–253. doi:10.1093/humupd/dmq037

Calvert ME, Digilio LC, Herr JC, Coonrod SA (2003) Oolemmal proteomics–identification of highly abundant heat shock proteins and molecular chaperones in the mature mouse egg and their localization on the plasma membrane. Reprod Biol Endocrinol (RB&E) 1:27

Camargo M, Intasqui P, Del Giudice PT, Carvalho VM, Cardozo KH, Andreoni C, Fraietta R, Bertolla RP (2013) Unbiased label-free quantitative proteomic profiling and enriched proteomic pathways in seminal plasma of adult men before and after varicocelectomy. Hum Reprod 28(1):33–46. doi:10.1093/humrep/des357

Camargo M, Intasqui P, de Lima CB, Montani DA, Nichi M, Pilau EJ, Gozzo FC, Lo Turco EG, Bertolla RP (2014) Maldi-tof fingerprinting of seminal plasma lipids in the study of human male infertility. Lipids 49(9):943–956. doi:10.1007/s11745-014-3922-7

Cao S, Guo X, Zhou Z, Sha J (2012) Comparative proteomic analysis of proteins involved in oocyte meiotic maturation in mice. Mol Reprod Dev 79(6):413–422. doi:10.1002/mrd.22044

Carlsen E, Giwercman A, Keiding N, Skakkebaek NE (1992) Evidence for decreasing quality of semen during past 50 years. BMJ (Clin Res ed) 305(6854):609–613

Carrell DT, Aston KI, Oliva R, Emery BR, De Jonge CJ (2016) The "omics" of human male infertility: integrating big data in a systems biology approach. Cell Tissue Res 363(1):295–312. doi:10.1007/s00441-015-2320-7

Carson DD, Bagchi I, Dey SK, Enders AC, Fazleabas AT, Lessey BA, Yoshinaga K (2000) Embryo implantation. Dev Biol 223(2):217–237. doi:10.1006/dbio.2000.9767

Casado-Vela J, Rodriguez-Suarez E, Iloro I, Ametzazurra A, Alkorta N, Garcia-Velasco JA, Matorras R, Prieto B, Gonzalez S, Nagore D, Simon L, Elortza F (2009) Comprehensive proteomic analysis of human endometrial fluid aspirate. J Proteome Res 8(10):4622–4632. doi:10.1021/pr9004426

Castillo J, Amaral A, Oliva R (2014) Sperm nuclear proteome and its epigenetic potential. Andrology 2(3):326–338. doi:10.1111/j.2047-2927.2013.00170.x

Cataldi T, Cordeiro FB, Costa Ldo V, Pilau EJ, Ferreira CR, Gozzo FC, Eberlin MN, Bertolla RP, Cedenho AP, Turco EG (2013) Lipid profiling of follicular fluid from women undergoing IVF: young poor ovarian responders versus normal responders. Hum Fertil (Cambridge, England) 16(4):269-277. doi:10.3109/14647273.2013.852255

Cervero A, Horcajadas JA, Dominguez F, Pellicer A, Simon C (2005) Leptin system in embryo development and implantation: a protein in search of a function. Reprod Biomed Online 10 (2):217–223

Chahrour O, Cobice D, Malone J (2015) Stable isotope labelling methods in mass spectrometry-based quantitative proteomics. J Pharm Biomed Anal 113:2–20. doi:10.1016/j. jpba.2015.04.013

Chan CC, Shui HA, Wu CH, Wang CY, Sun GH, Chen HM, Wu GJ (2009) Motility and protein phosphorylation in healthy and asthenozoospermic sperm. J Proteome Res 8(11):5382–5386. doi:10.1021/pr9003932

Chandramouli K, Qian PY (2009) Proteomics: challenges, techniques and possibilities to overcome biological sample complexity. Hum Genomics Proteomics: HGP 2009. doi:10.4061/ 2009/239204

Chen Q, Zhang A, Yu F, Gao J, Liu Y, Yu C, Zhou H, Xu C (2015) Label-free proteomics uncovers energy metabolism and focal adhesion regulations responsive for endometrium receptivity. J Proteome Res 14(4):1831–1842. doi:10.1021/acs.jproteome.5b00038

Chiasserini D, Mazzoni M, Bordi F, Sennato S, Susta F, Orvietani PL, Binaglia L, Palmerini CA (2015) Identification and partial characterization of two populations of prostasomes by a combination of dynamic light scattering and proteomic analysis. J Membr Biol 248(6):991–1004. doi:10.1007/s00232-015-9810-0

Cho WC (2007) Contribution of oncoproteomics to cancer biomarker discovery. Mol Cancer 6:25. doi:10.1186/1476-4598-6-25

Choe L, D'Ascenzo M, Relkin NR, Pappin D, Ross P, Williamson B, Guertin S, Pribil P, Lee KH (2007) 8-plex quantitation of changes in cerebrospinal fluid protein expression in subjects undergoing intravenous immunoglobulin treatment for Alzheimer's disease. Proteomics 7 (20):3651–3660. doi:10.1002/pmic.200700316

Choi WS, Kim SW (2013) Current issues in varicocele management: a review. World J Men's Health 31(1):12–20. doi:10.5534/wjmh.2013.31.1.12

Chong PK, Gan CS, Pham TK, Wright PC (2006) Isobaric tags for relative and absolute quantitation (iTRAQ) reproducibility: Implication of multiple injections. J Proteome Res 5 (5):1232–1240. doi:10.1021/pr060018u

Chretien FC (2003) Involvement of the glycoproteic meshwork of cervical mucus in the mechanism of sperm orientation. Acta obstetricia et gynecologica Scandinavica 82(5):449–461

Claassen M (2012) Inference and validation of protein identifications. Mol Cell Proteomics MCP 11(11):1097–1104. doi:10.1074/mcp.R111.014795

Clamp M, Fry B, Kamal M, Xie X, Cuff J, Lin MF, Kellis M, Lindblad-Toh K, Lander ES (2007) Distinguishing protein-coding and noncoding genes in the human genome. Proc Natl Acad Sci U.S.A 104(49):19428–19433. doi:10.1073/pnas.0709013104

Clifton J, Huang F, Rucevic M, Cao L, Hixson D, Josic D (2011) Protease inhibitors as possible pitfalls in proteomic analyses of complex biological samples. J Proteomics 74(7):935–941. doi:10.1016/j.jprot.2011.02.010

Cline MS, Smoot M, Cerami E, Kuchinsky A, Landys N, Workman C, Christmas R, Avila-Campilo I, Creech M, Gross B, Hanspers K, Isserlin R, Kelley R, Killcoyne S, Lotia S, Maere S, Morris J, Ono K, Pavlovic V, Pico AR, Vailaya A, Wang PL, Adler A, Conklin BR, Hood L, Kuiper M, Sander C, Schmulevich I, Schwikowski B, Warner GJ, Ideker T, Bader GD (2007) Integration of biological networks and gene expression data using Cytoscape. Nat Protoc 2(10):2366–2382. doi:10.1038/nprot.2007.324

Cocuzza M, Cocuzza MA, Bragais FM, Agarwal A (2008) The role of varicocele repair in the new era of assisted reproductive technology. Clinics 63(3):395–404

Cornwall GA (2009) New insights into epididymal biology and function. Hum Reprod Update 15 (2):213–227. doi:10.1093/humupd/dmn055

Cortezzi SS, Garcia JS, Ferreira CR, Braga DP, Figueira RC, Iaconelli A Jr, Souza GH, Borges E Jr, Eberlin MN (2011) Secretome of the preimplantation human embryo by bottom-up label-free proteomics. Anal Bioanal Chem 401(4):1331–1339. doi:10.1007/s00216-011-5202-1

Corthals GL, Wasinger VC, Hochstrasser DF, Sanchez JC (2000) The dynamic range of protein expression: a challenge for proteomic research. Electrophoresis 21(6):1104–1115. doi:10.1002/ (SICI)1522-2683(20000401)21:6<1104:AID-ELPS1104>3.0.CO;2-C

Coughlan C, Ledger W, Wang Q, Liu F, Demirol A, Gurgan T, Cutting R, Ong K, Sallam H, Li TC (2014) Recurrent implantation failure: definition and management. Reprod Biomed Online 28(1):14–38. doi:10.1016/j.rbmo.2013.08.011

Cox J, Mann M (2008) MaxQuant enables high peptide identification rates, individualized p.p.b.-range mass accuracies and proteome-wide protein quantification. Nat Biotechnol 26(12):1367–1372. doi:10.1038/nbt.1511

Cox J, Mann M (2011) Quantitative, high-resolution proteomics for data-driven systems biology. Ann Rev Biochem 80:273–299. doi:10.1146/annurev-biochem-061308-093216

Cozzolino DJ, Lipshultz LI (2001) Varicocele as a progressive lesion: positive effect of varicocele repair. Hum Reprod Update 7(1):55–58

Croft D, Mundo AF, Haw R, Milacic M, Weiser J, Wu G, Caudy M, Garapati P, Gillespie M, Kamdar MR, Jassal B, Jupe S, Matthews L, May B, Palatnik S, Rothfels K, Shamovsky V, Song H, Williams M, Birney E, Hermjakob H, Stein L, D'Eustachio P (2014) The reactome pathway knowledgebase. Nucleic Acids Res 42(Database issue):D472–D477. doi:10.1093/nar/ gkt1102

D'Amours O, Bordeleau LJ, Frenette G, Blondin P, Leclerc P, Sullivan R (2012) Binder of sperm 1 and epididymal sperm binding protein 1 are associated with different bull sperm subpopulations. Reproduction 143(6):759–771. doi:10.1530/rep-11-0392

da Silva BF, Souza GH, lo Turco EG, Del Giudice PT, Soler TB, Spaine DM, Borrelli Junior M, Gozzo FC, Pilau EJ, Garcia JS, Ferreira CR, Eberlin MN, Bertolla RP (2013) Differential seminal plasma proteome according to semen retrieval in men with spinal cord injury. Fertil Steril 100(4):959-969. doi:10.1016/j.fertnstert.2013.06.009

da Silva BF, Meng C, Helm D, Pachl F, Schiller J, Ibrahim E, Lynne CM, Brackett NL, Bertolla RP, Kuster B (2016) Towards understanding male infertility after spinal cord injury using quantitative proteomics. Mol Cell Proteomics MCP 15(4):1424–1434. doi:10.1074/mcp. M115.052175

da Silveira JC, Veeramachaneni DN, Winger QA, Carnevale EM, Bouma GJ (2012) Cell-secreted vesicles in equine ovarian follicular fluid contain miRNAs and proteins: a possible new form of cell communication within the ovarian follicle. Biol Reprod 86(3):71. doi:10.1095/biolreprod. 111.093252

da Silveira JC, Winger QA, Bouma GJ, Carnevale EM (2015) Effects of age on follicular fluid exosomal microRNAs and granulosa cell transforming growth factor-beta signalling during follicle development in the mare. Reprod Fertil Dev 27(6):897–905. doi:10.1071/rd14452

Dacheux JL, Dacheux F (2014) New insights into epididymal function in relation to sperm maturation. Reproduction 147(2):R27–R42. doi:10.1530/REP-13-0420

Dai G, Lu G (2012) Different protein expression patterns associated with polycystic ovary syndrome in human follicular fluid during controlled ovarian hyperstimulation. Reprod Fertil Dev 24(7):893–904. doi:10.1071/RD11201

Dasari S, Pereira L, Reddy AP, Michaels JE, Lu X, Jacob T, Thomas A, Rodland M, Roberts CT Jr, Gravett MG, Nagalla SR (2007) Comprehensive proteomic analysis of human cervical-vaginal fluid. J Proteome Res 6(4):1258–1268. doi:10.1021/pr0605419

Davalieva K, Kiprijanovska S, Noveski P, Plaseski T, Kocevska B, Broussard C, Plaseska-Karanfilska D (2012) Proteomic analysis of seminal plasma in men with different spermatogenic impairment. Andrologia 44(4):256–264. doi:10.1111/j.1439-0272.2012.01275.x

de Kretser DM (1997) Male infertility. Lancet 349(9054):787–790

de Lamirande E, O'Flaherty C (2012) Sperm capacitation as an oxidative event. Studies on men's health and fertility, oxidative stress in applied basic research and clinical practice, Springer Science, pp 57–94 de Mateo S, Martinez-Heredia J, Estanyol JM, Dominguez-Fandos D, Vidal-Taboada JM, Ballesca JL, Oliva R (2007) Marked correlations in protein expression identified by proteomic analysis of human spermatozoa. Proteomics 7(23):4264–4277. doi:10.1002/pmic.200700521

de Mateo S, Castillo J, Estanyol JM, Ballesca JL, Oliva R (2011) Proteomic characterization of the human sperm nucleus. Proteomics 11(13):2714–2726. doi:10.1002/pmic.201000799

de Mouzon J, Goossens V, Bhattacharya S, Castilla JA, Ferraretti AP, Korsak V, Kupka M, Nygren KG, Nyboe Andersen A, European Ivf-monitoring Consortium ftESoHR, Embryology (2010) Assisted reproductive technology in Europe, 2006: results generated from European registers by ESHRE. Hum Reprod 25(8):1851–1862. doi:10.1093/humrep/deq124

De Oliveira L, Camara NO, Bonetti T, Lo Turco EG, Bertolla RP, Moron AF, Sass N, Da Silva ID (2012) Lipid fingerprinting in women with early-onset preeclampsia: a first look. Clinical Biochem 45(10–11):852–855. doi:10.1016/j.clinbiochem.2012.04.012

de Rond T, Danielewicz M, Northen T (2015) High throughput screening of enzyme activity with mass spectrometry imaging. Curr Opin Biotechnol 31:1–9. doi:10.1016/j.copbio.2014.07.008

Del Giudice PT, Lima SB, Cenedeze MA, Pacheco-Silva A, Bertolla RP, Cedenho AP (2010) Expression of the Fas-ligand gene in ejaculated sperm from adolescents with and without varicocele. J Assist Reproduct Genet 27(2–3):103–109. doi:10.1007/s10815-010-9384-9

Del Giudice PT, da Silva BF, Lo Turco EG, Fraietta R, Spaine DM, Santos LF, Pilau EJ, Gozzo FC, Cedenho AP, Bertolla RP (2013) Changes in the seminal plasma proteome of adolescents before and after varicocelectomy. Fertil Steril 100(3):667–672. doi:10.1016/j.fertnstert.2013.04.036

Del Giudice PT, Belardin LB, Camargo M, Zylbersztejn DS, Carvalho VM, Cardozo KH, Bertolla RP, Cedenho AP (2016) Determination of testicular function in adolescents with varicocoele—a proteomics approach. Andrology 4(3):447–455. doi:10.1111/andr.12174

Delmotte N, Lasaosa M, Tholey A, Heinzle E, Huber CG (2007) Two-dimensional reversed-phase x ion-pair reversed-phase HPLC: an alternative approach to high-resolution peptide separation for shotgun proteome analysis. J Proteome Res 6(11):4363–4373. doi:10.1021/pr070424t

DeSouza L, Diehl G, Yang EC, Guo J, Rodrigues MJ, Romaschin AD, Colgan TJ, Siu KW (2005) Proteomic analysis of the proliferative and secretory phases of the human endometrium: protein identification and differential protein expression. Proteomics 5(1):270–281. doi:10.1002/pmic.200400920

Di Quinzio MK, Oliva K, Holdsworth SJ, Ayhan M, Walker SP, Rice GE, Georgiou HM, Permezel M (2007) Proteomic analysis and characterisation of human cervico-vaginal fluid proteins. Aust NZ J Obstet Gynaecol 47(1):9–15. doi:10.1111/j.1479-828X.2006.00671.x

Di Quinzio MK, Georgiou HM, Holdsworth-Carson SJ, Ayhan M, Heng YJ, Walker SP, Rice GE, Permezel M (2008) Proteomic analysis of human cervico-vaginal fluid displays differential protein expression in association with labor onset at term. J Proteome Res 7(5):1916–1921. doi:10.1021/pr7006413

Diaz-Cueto L, Stein P, Jacobs A, Schultz RM, Gerton GL (2000) Modulation of mouse preimplantation embryo development by acrogranin (epithelin/granulin precursor). Dev Biol 217(2):406–418. doi:10.1006/dbio.1999.9564

Dominguez F, Gadea B, Esteban FJ, Horcajadas JA, Pellicer A, Simon C (2008) Comparative protein-profile analysis of implanted versus non-implanted human blastocysts. Hum Reprod 23(9):1993–2000. doi:10.1093/humrep/den205

Dominguez F, Garrido-Gomez T, Lopez JA, Camafeita E, Quinonero A, Pellicer A, Simon C (2009) Proteomic analysis of the human receptive versus non-receptive endometrium using differential in-gel electrophoresis and MALDI-MS unveils stathmin 1 and annexin A2 as differentially regulated. Hum Reprod 24(10):2607–2617. doi:10.1093/humrep/dep230

Drabovich AP, Jarvi K, Diamandis EP (2011) Verification of male infertility biomarkers in seminal plasma by multiplex selected reaction monitoring assay. Mol Cell Proteomics: MCP 10(12):M110 004127. doi:10.1074/mcp.M110.004127

Drabovich AP, Dimitromanolakis A, Saraon P, Soosaipillai A, Batruch I, Mullen B, Jarvi K, Diamandis EP (2013) Differential diagnosis of azoospermia with proteomic biomarkers ECM1 and TEX101 quantified in seminal plasma. Sci Transl Med 5(212):212ra160. doi:10.1126/scitranslmed.3006260

Drabovich AP, Saraon P, Jarvi K, Diamandis EP (2014) Seminal plasma as a diagnostic fluid for male reproductive system disorders. Nat Rev Urol 11(5):278–288. doi:10.1038/nrurol.2014.74

Druart X (2012) Sperm interaction with the female reproductive tract. Reproduction in domestic animals = Zuchthygiene 47(Suppl 4):348–352. doi:10.1111/j.1439-0531.2012.02097.x

Dunson DB, Baird DD, Colombo B (2004) Increased infertility with age in men and women. Obstet Gynecol 103(1):51–56. doi:10.1097/01.AOG.0000100153.24061.45

Dutta M, Subramani E, Taunk K, Gajbhiye A, Seal S, Pendharkar N, Dhali S, Ray CD, Lodh I, Chakravarty B, Dasgupta S, Rapole S, Chaudhury K (2015) Investigation of serum proteome alterations in human endometriosis. J Proteomics 114:182–196. doi:10.1016/j.jprot.2014.10.021

Dyrlund TF, Kirkegaard K, Poulsen ET, Sanggaard KW, Hindkjaer JJ, Kjems J, Enghild JJ, Ingerslev HJ (2014) Unconditioned commercial embryo culture media contain a large variety of non-declared proteins: a comprehensive proteomics analysis. Hum Reprod 29(11):2421–2430. doi:10.1093/humrep/deu220

Egea RR, Puchalt NG, Escriva MM, Varghese AC (2014) OMICS: Current and future perspectives in reproductive medicine and technology. J Hum Reprod Sci 7(2):73–92. doi:10.4103/0974-1208.138857

Erikson DW, Way AL, Bertolla RP, Chapman DA, Killian GJ (2007a) Influence of osteopontin, casein and oviductal fluid on bovine sperm capacitation. Anim Reprod 4(3/4):103–112

Erikson DW, Way AL, Chapman DA, Killian GJ (2007b) Detection of osteopontin on Holstein bull spermatozoa, in cauda epididymal fluid and testis homogenates, and its potential role in bovine fertilization. Reproduction 133(5):909–917. doi:10.1530/REP-06-0228

Estes SJ, Ye B, Qiu W, Cramer D, Hornstein MD, Missmer SA (2009) A proteomic analysis of IVF follicular fluid in women <or= 32 years old. Fertil Steril 92(5):1569–1578. doi:10.1016/j.fertnstert.2008.08.120

Esteves SC, Agarwal A (2013) The azoospermic male: current knowledge and future perspectives. Clinics 68(Suppl 1):1–4

Esteves SC, Agarwal A (2016) Afterword to varicocele and male infertility: current concepts and future perspectives. Asian J Androl 18(2):319–322. doi:10.4103/1008-682X.172820

Etzioni R, Kooperberg C, Pepe M, Smith R, Gann PH (2003) Combining biomarkers to detect disease with application to prostate cancer. Biostatistics (Oxford, England) 4(4):523–538. doi:10.1093/biostatistics/4.4.523

Fariello RM, Pariz JR, Spaine DM, Gozzo FC, Pilau EJ, Fraietta R, Bertolla RP, Andreoni C, Cedenho AP (2012) Effect of smoking on the functional aspects of sperm and seminal plasma protein profiles in patients with varicocele. Hum Reprod 27(11):3140–3149. doi:10.1093/humrep/des287

Fassbender A, Verbeeck N, Bornigen D, Kyama CM, Bokor A, Vodolazkaia A, Peeraer K, Tomassetti C, Meuleman C, Gevaert O, Van de Plas R, Ojeda F, De Moor B, Moreau Y, Waelkens E, D'Hooghe TM (2012) Combined mRNA microarray and proteomic analysis of eutopic endometrium of women with and without endometriosis. Hum Reprod 27(7):2020–2029. doi:10.1093/humrep/des127

Feist P, Hummon AB (2015) Proteomic challenges: sample preparation techniques for microgram-quantity protein analysis from biological samples. Int J Mol Sci 16(2):3537–3563. doi:10.3390/ijms16023537

Fenn JB, Mann M, Meng CK, Wong SF, Whitehouse CM (1989) Electrospray ionization for mass spectrometry of large biomolecules. Science 246(4926):64–71

Feng S, Ye M, Zhou H, Jiang X, Jiang X, Zou H, Gong B (2007) Immobilized zirconium ion affinity chromatography for specific enrichment of phosphopeptides in phosphoproteome analysis. Mol Cell Proteomics MCP 6(9):1656–1665. doi:10.1074/mcp.T600071-MCP200

Ferreira C, Lo Turco R, Saraiva S, Bertolla R, Perecin F, Souza G, Murgu M, Garcia J, Cortezzi S, Meirelles F (2010) Proteomics, metabolomis and lipidomics in reproductive biotechnologies: the MS solutions. Acta Scientiae Veterinariae 38(Suppl. 2):s591–s603

Ferraretti AP, Goossens V, de Mouzon J, Bhattacharya S, Castilla JA, Korsak V, Kupka M, Nygren KG, Nyboe Andersen A, European IVFm, Consortium for European Society of Human R, Embryology (2012) Assisted reproductive technology in Europe, 2008: results generated from European registers by ESHRE. Hum Reprod 27(9):2571–2584. doi:10.1093/humrep/des255

Ferrero S, Gillott DJ, Remorgida V, Anserini P, Leung KY, Ragni N, Grudzinskas JG (2007) Proteomic analysis of peritoneal fluid in women with endometriosis. J Proteome Res 6 (9):3402–3411. doi:10.1021/pr060680q

Ferrero S, Gillott DJ, Remorgida V, Anserini P, Ragni N, Grudzinskas JG (2009) Proteomic analysis of peritoneal fluid in fertile and infertile women with endometriosis. J Reprod Med 54 (1):32–40

Fraser LR (1998) Sperm capacitation and the acrosome reaction. Hum Reprod 13(Suppl 1):9–19

Freour T, Com E, Barriere P, Bouchot O, Jean M, Masson D, Pineau C (2013) Comparative proteomic analysis coupled with conventional protein assay as a strategy to identify predictors of successful testicular sperm extraction in patients with non-obstructive azoospermia. Andrology 1(3):414–420. doi:10.1111/j.2047-2927.2012.00059.x

Fowler PA, Tattum J, Bhattacharya S, Klonisch T, Hombach-Klonisch S, Gazvani R, Lea RG, Miller I, Simpson WG, Cash P (2007) An investigation of the effects of endometriosis on the proteome of human eutopic endometrium: a heterogeneous tissue with a complex disease. Proteomics 7(1):130–142. doi:10.1002/pmic.200600469

Fu-Jun L, Xiao-Fang S (2012) Comparative analysis of human reproductive proteomes identifies candidate proteins of sperm maturation. Mol Biol Rep 39(12):10257–10263. doi:10.1007/s11033-012-1902-7

Fu Y, Qian X (2014) Transferred subgroup false discovery rate for rare post-translational modifications detected by mass spectrometry. Mol Cell Proteomics MCP 13(5):1359–1368. doi:10.1074/mcp.O113.030189

Fuhrer T, Zamboni N (2015) High-throughput discovery metabolomics. Current Opin Biotechnol 31:73–78. doi:10.1016/j.copbio.2014.08.006

Fung KY, Glode LM, Green S, Duncan MW (2004) A comprehensive characterization of the peptide and protein constituents of human seminal fluid. Prostate 61(2):171–181. doi:10.1002/pros.20089

Fuzery AK, Levin J, Chan MM, Chan DW (2013) Translation of proteomic biomarkers into FDA approved cancer diagnostics: issues and challenges. Clin Proteomics 10(1):13. doi:10.1186/1559-0275-10-13

Galazis N, Olaleye O, Haoula Z, Layfield R, Atiomo W (2012) Proteomic biomarkers for ovarian cancer risk in women with polycystic ovary syndrome: a systematic review and biomarker database integration. Fertil Steril 98(6):1590–1601 e1591. doi:10.1016/j.fertnstert.2012.08.002

Galazis N, Pang YL, Galazi M, Haoula Z, Layfield R, Atiomo W (2013) Proteomic biomarkers of endometrial cancer risk in women with polycystic ovary syndrome: a systematic review and biomarker database integration. Gynecol Endocrinol 29(7):638–644. doi:10.3109/09513590.2013.777416

Gallien S, Kim SY, Domon B (2015) Large-scale targeted proteomics using internal standard triggered-parallel reaction monitoring (IS-PRM). Mol Cell Proteomics MCP 14(6):1630–1644. doi:10.1074/mcp.O114.043968

Gao L, Song Q, Patterson GE, Cooks RG, Ouyang Z (2006) Handheld rectilinear ion trap mass spectrometer. Anal Chem 78(17):5994–6002. doi:10.1021/ac061144k

Gao J, Tarcea VG, Karnovsky A, Mirel BR, Weymouth TE, Beecher CW, Cavalcoli JD, Athey BD, Omenn GS, Burant CF, Jagadish HV (2010) Metscape: a Cytoscape plug-in for visualizing and interpreting metabolomic data in the context of human metabolic networks. Bioi (Oxford, England) 26(7):971–973. doi:10.1093/bioinformatics/btq048

Gardner DK, Meseguer M, Rubio C, Treff NR (2015) Diagnosis of human preimplantation embryo viability. Hum Reprod Update 21(6):727–747. doi:10.1093/humupd/dmu064

Garrido-Gomez T, Dominguez F, Lopez JA, Camafeita E, Quinonero A, Martinez-Conejero JA, Pellicer A, Conesa A, Simon C (2011) Modeling human endometrial decidualization from the interaction between proteome and secretome. J Clin Endocrinol Metab 96(3):706–716. doi:10.1210/jc.2010-1825

Garrido-Gomez T, Quinonero A, Antunez O, Diaz-Gimeno P, Bellver J, Simon C, Dominguez F (2014) Deciphering the proteomic signature of human endometrial receptivity. Hum Reprod 29 (9):1957–1967. doi:10.1093/humrep/deu171

Geoffroy-Siraudin C, Loundou AD, Romain F, Achard V, Courbiere B, Perrard MH, Durand P, Guichaoua MR (2012) Decline of semen quality among 10 932 males consulting for couple infertility over a 20-year period in Marseille, France. Asian J Androl 14(4):584–590. doi:10.1038/aja.2011.173

Giacomini E, Ura B, Giolo E, Luppi S, Martinelli M, Garcia RC, Ricci G (2015) Comparative analysis of the seminal plasma proteomes of oligoasthenozoospermic and normozoospermic men. Reprod Biomed Online 30(5):522–531. doi:10.1016/j.rbmo.2015.01.010

Gianaroli L, Magli MC, Cavallini G, Crippa A, Capoti A, Resta S, Robles F, Ferraretti AP (2010) Predicting aneuploidy in human oocytes: key factors which affect the meiotic process. Hum Reprod 25(9):2374–2386. doi:10.1093/humrep/deq123

Giannopoulou EG, Garbis SD, Vlahou A, Kossida S, Lepouras G, Manolakos ES (2009) Proteomic feature maps: a new visualization approach in proteomics analysis. J Biomed Inform 42(4):644–653. doi:10.1016/j.jbi.2009.01.007

Glander HJ, Schiller J, Suss R, Paasch U, Grunewald S, Arnhold J (2002) Deterioration of spermatozoal plasma membrane is associated with an increase of sperm lyso-phosphatidylcholines. Andrologia 34(6):360–366

Gonzalez RR, Caballero-Campo P, Jasper M, Mercader A, Devoto L, Pellicer A, Simon C (2000) Leptin and leptin receptor are expressed in the human endometrium and endometrial leptin secretion is regulated by the human blastocyst. J Clin Endocrinol Metab 85(12):4883–4888. doi:10.1210/jcem.85.12.7060

Gorelick JI, Goldstein M (1993) Loss of fertility in men with varicocele. Fertil Steril 59(3):613–616

Goshe MB, Conrads TP, Panisko EA, Angell NH, Veenstra TD, Smith RD (2001) Phosphoprotein isotope-coded affinity tag approach for isolating and quantitating phosphopeptides in proteome-wide analyses. Anal Chem 73(11):2578–2586

Grande G, Milardi D, Vincenzoni F, Pompa G, Biscione A, Astorri AL, Fruscella E, De Luca A, Messana I, Castagnola M, Marana R (2015) Proteomic characterization of the qualitative and quantitative differences in cervical mucus composition during the menstrual cycle. Mol BioSyst 11(6):1717–1725. doi:10.1039/c5mb00071h

Gupta S, Sharma R, Eliwa J, Agarwal A (2015) Predictive roles of proteomic profiles in assisted reproduction-an update. J Proteomics Bioinf. doi: doi: 10.4172/jpb.S8-005

Gutstein HB, Morris JS, Annangudi SP, Sweedler JV (2008) Microproteomics: analysis of protein diversity in small samples. Mass Spectrome Rev 27(4):316–330. doi:10.1002/mas.20161

Haga SW, Wu HF (2014) Overview of software options for processing, analysis and interpretation of mass spectrometric proteomic data. J Mass Spectrom JMS 49(10):959–969. doi:10.1002/jms.3414

Haimov-Kochman R, Har-Nir R, Ein-Mor E, Ben-Shoshan V, Greenfield C, Eldar I, Bdolah Y, Hurwitz A (2012) Is the quality of donated semen deteriorating? Findings from a 15 year longitudinal analysis of weekly sperm samples. Israel Med Assoc J IMAJ 14(6):372–377

Hamada A, Sharma R, du Plessis SS, Willard B, Yadav SP, Sabanegh E, Agarwal A (2013) Two-dimensional differential in-gel electrophoresis-based proteomics of male gametes in relation to oxidative stress. Fertil Steril 99(5):1216–1226 e1212. doi:10.1016/j.fertnstert.2012.11.046

Hamamah S, Matha V, Berthenet C, Anahory T, Loup V, Dechaud H, Hedon B, Fernandez A, Lamb N (2006) Comparative protein expression profiling in human cumulus cells in relation to oocyte fertilization and ovarian stimulation protocol. Reprod Biomed Online 13(6):807–814

Hannan NJ, Stephens AN, Rainczuk A, Hincks C, Rombauts LJ, Salamonsen LA (2010) 2D-DiGE analysis of the human endometrial secretome reveals differences between receptive and nonreceptive states in fertile and infertile women. J Proteome Res 9(12):6256–6264. doi:10.1021/pr1004828

Hanrieder J, Nyakas A, Naessen T, Bergquist J (2008) Proteomic analysis of human follicular fluid using an alternative bottom-up approach. J Proteome Res 7(1):443–449. doi:10.1021/pr070277z

Hartl FU, Hayer-Hartl M (2002) Molecular chaperones in the cytosol: from nascent chain to folded protein. Science 295(5561):1852–1858. doi:10.1126/science.1068408

Hartl FU, Hayer-Hartl M (2009) Converging concepts of protein folding in vitro and in vivo. Nat Struct Mol Biol 16(6):574–581. doi:10.1038/nsmb.1591

Hashemitabar M, Bahmanzadeh M, Mostafaie A, Orazizadeh M, Farimani M, Nikbakht R (2014) A proteomic analysis of human follicular fluid: comparison between younger and older women with normal FSH levels. Int J Mol Sci 15(10):17518–17540. doi:10.3390/ijms151017518

Hashemitabar M, Sabbagh S, Orazizadeh M, Ghadiri A, Bahmanzadeh M (2015) A proteomic analysis on human sperm tail: comparison between normozoospermia and asthenozoospermia. J Assist Reprod Genet 32(6):853–863. doi:10.1007/s10815-015-0465-7

Hassan MA, Killick SR (2003) Effect of male age on fertility: evidence for the decline in male fertility with increasing age. Fertil Steril 79(Suppl 3):1520–1527

Heeren RM, Smith DF, Stauber J, Kukrer-Kaletas B, MacAleese L (2009) Imaging mass spectrometry: hype or hope? J Am Soc Mass Spectrom 20(6):1006–1014. doi:10.1016/j.jasms.2009.01.011

Heng YJ, Di Quinzio MK, Permezel M, Ayhan M, Rice GE, Georgiou HM (2010) Temporal proteomic analysis of human cervicovaginal fluid with impending term labor. J Proteome Res 9(3):1344–1350. doi:10.1021/pr900892f

Heng YJ, Di Quinzio MK, Permezel M, Rice GE, Georgiou HM (2008) Interleukin-1 receptor antagonist in human cervicovaginal fluid in term pregnancy and labor. Am J Obstet Gynecol 199(6):656 e651–e657. doi:10.1016/j.ajog.2008.06.011

Hermo L, Pelletier RM, Cyr DG, Smith CE (2010a) Surfing the wave, cycle, life history, and genes/proteins expressed by testicular germ cells. Part 1: background to spermatogenesis, spermatogonia, and spermatocytes. Microsc Res Tech 73(4):241–278. doi:10.1002/jemt.20783

Hermo L, Pelletier RM, Cyr DG, Smith CE (2010b) Surfing the wave, cycle, life history, and genes/proteins expressed by testicular germ cells. Part 5: intercellular junctions and contacts between germs cells and Sertoli cells and their regulatory interactions, testicular cholesterol, and genes/proteins associated with more than one germ cell generation. Microsc Res Tech 73(4):409–494. doi:10.1002/jemt.20786

Herwig R, Knoll C, Planyavsky M, Pourbiabany A, Greilberger J, Bennett KL (2013) Proteomic analysis of seminal plasma from infertile patients with oligoasthenoteratozoospermia due to oxidative stress and comparison with fertile volunteers. Fertil Steri 100(2):355–366 e352. doi:10.1016/j.fertnstert.2013.03.048

Hicks WA, Halligan BD, Slyper RY, Twigger SN, Greene AS, Olivier M (2005) Simultaneous quantification and identification using 18O labeling with an ion trap mass spectrometer and the analysis software application "ZoomQuant". J Am Soc Mass Spectrom 16(6):916–925. doi:10.1016/j.jasms.2005.02.024

Hoedt E, Zhang G, Neubert TA (2014) Stable isotope labeling by amino acids in cell culture (SILAC) for quantitative proteomics. Adv Exp Med Biol 806:93–106. doi:10.1007/978-3-319-06068-2_5

Homa ST, Vessey W, Perez-Miranda A, Riyait T, Agarwal A (2015) Reactive Oxygen Species (ROS) in human semen: determination of a reference range. J Assist Reprod Genet 32(5):757–764. doi:10.1007/s10815-015-0454-x

Hood L, Flores M (2012) A personal view on systems medicine and the emergence of proactive P4 medicine: predictive, preventive, personalized and participatory. New Biotechnol 29(6):613–624. doi:10.1016/j.nbt.2012.03.004

Hood BL, Liu B, Alkhas A, Shoji Y, Challa R, Wang G, Ferguson S, Oliver J, Mitchell D, Bateman NW, Zahn CM, Hamilton CA, Payson M, Lessey B, Fazleabas AT, Maxwell GL, Conrads TP, Risinger JI (2015) Proteomics of the human endometrial glandular epithelium and stroma from the proliferative and secretory phases of the menstrual cycle. Biol Reprod 92 (4):106. doi:10.1095/biolreprod.114.127217

Hosseinifar H, Sabbaghian M, Nasrabadi D, Modarresi T, Dizaj AV, Gourabi H, Gilani MA (2014) Study of the effect of varicocelectomy on sperm proteins expression in patients with varicocele and poor sperm quality by using two-dimensional gel electrophoresis. J Assist Reprod Genet 31(6):725–729. doi:10.1007/s10815-014-0209-0

Hsu CC, Dorrestein PC (2015) Visualizing life with ambient mass spectrometry. Curr Opin Biotechnol 31:24–34. doi:10.1016/j.copbio.2014.07.005

Hulka BS (1990) Overview of biological markers. In: Hulka BSGJ, Wilcosky TC (eds) Biological markers in epidemiology. Oxford University Press, New York, pp 3–15

Huntriss J, Hinkins M, Picton HM (2006) cDNA cloning and expression of the human NOBOX gene in oocytes and ovarian follicles. Mol Hum Reprod 12(5):283–289. doi:10.1093/molehr/gal035

Hwang JH, Oh JJ, Wang T, Jin YC, Lee JS, Choi JR, Lee KS, Joo JK, Lee HG (2013) Identification of biomarkers for endometriosis in eutopic endometrial cells from patients with endometriosis using a proteomics approach. Mol Med Rep 8(1):183–188. doi:10.3892/mmr.2013.1469

Hwang JH, Lee KS, Joo JK, Wang T, Son JB, Park JH, Hwang DY, Choi MH, Lee HG (2014) Identification of biomarkers for endometriosis in plasma from patients with endometriosis using a proteomics approach. Mo Med Rep 10(2):725–730. doi:10.3892/mmr.2014.2291

Ignatoski KM (2001) Immunoprecipitation and western blotting of phosphotyrosine-containing proteins. Methods Mol Biol (Clifton, NJ) 124:39–48

Insenser M, Escobar-Morreale HF (2011) Application of proteomics to the study of polycystic ovary syndrome. J Endocrinol Invest 34(11):869–875. doi:10.3275/8108

Insenser M, Martinez-Garcia MA, Montes R, San-Millan JL, Escobar-Morreale HF (2010) Proteomic analysis of plasma in the polycystic ovary syndrome identifies novel markers involved in iron metabolism, acute-phase response, and inflammation. J Clin Endocrinol Metab 95(8):3863–3870. doi:10.1210/jc.2010-0220

Insenser M, Montes-Nieto R, Murri M, Escobar-Morreale HF (2013) Proteomic and metabolomic approaches to the study of polycystic ovary syndrome. Mol Cell Endocrinol 370(1–2):65–77. doi:10.1016/j.mce.2013.02.009

Intasqui P, Camargo M, Del Giudice PT, Spaine DM, Carvalho VM, Cardozo KH, Cedenho AP, Bertolla RP (2013a) Unraveling the sperm proteome and post-genomic pathways associated with sperm nuclear DNA fragmentation. J Assist Reprod Genet 30(9):1187–1202. doi:10.1007/s10815-013-0054-6

Intasqui P, Camargo M, Del Giudice PT, Spaine DM, Carvalho VM, Cardozo KH, Zylbersztejn DS, Bertolla RP (2013b) Sperm nuclear DNA fragmentation rate is associated with differential protein expression and enriched functions in human seminal plasma. BJU Int 112(6):835–843. doi:10.1111/bju.12233

Intasqui P, Antoniassi MP, Camargo M, Nichi M, Carvalho VM, Cardozo KH, Zylbersztejn DS, Bertolla RP (2015) Differences in the seminal plasma proteome are associated with oxidative stress levels in men with normal semen parameters. Fertil Steril 104(2):292–301. doi:10.1016/j.fertnstert.2015.04.037

Intasqui P, Camargo M, Antoniassi MP, Cedenho AP, Carvalho VM, Cardozo KH, Zylbersztejn DS, Bertolla RP (2016) Association between the seminal plasma proteome and sperm functional traits. Fertil Steril 105(3):617–628. doi:10.1016/j.fertnstert.2015.11.005

Irvine S, Cawood E, Richardson D, MacDonald E, Aitken J (1996) Evidence of deteriorating semen quality in the United Kingdom: birth cohort study in 577 men in Scotland over 11 years. BMJ (Clin Res ed) 312(7029):467–471

Izzo CR, Monteleone PA (1992) Serafini PC (2015) human reproduction: current status. Revista da Associacao Medica Brasileira 61(6):557–559. doi:10.1590/1806-9282.61.06.557

James P (1997) Protein identification in the post-genome era: the rapid rise of proteomics. Q Rev Biophys 30(4):279–331

Jensen SS, Larsen MR (2007) Evaluation of the impact of some experimental procedures on different phosphopeptide enrichment techniques. Rapid Commun Mass Spectrom RCM 21 (22):3635–3645. doi:10.1002/rcm.3254

Jansen C, Hebeda KM, Linkels M, Grefte JM, Raemaekers JM, van Krieken JH, Groenen PJ (2008) Protein profiling of B-cell lymphomas using tissue biopsies: a potential tool for small samples in pathology. Cell oncology Official J Int Soc Cell Oncol 30(1):27–38

Jarkovska K, Martinkova J, Liskova L, Halada P, Moos J, Rezabek K, Gadher SJ, Kovarova H (2010) Proteome mining of human follicular fluid reveals a crucial role of complement cascade and key biological pathways in women undergoing in vitro fertilization. J Proteome Res 9 (3):1289–1301. doi:10.1021/pr900802u

Jeong H, Mason SP, Barabasi AL, Oltvai ZN (2001) Lethality and centrality in protein networks. Nature 411(6833):41–42. doi:10.1038/35075138

Johnston DS, Wooters J, Kopf GS, Qiu Y, Roberts KP (2005) Analysis of the human sperm proteome. Ann NY Acad Sci 1061:190–202. doi:10.1196/annals.1336.021

Junger MA, Aebersold R (2014) Mass spectrometry-driven phosphoproteomics: patterning the systems biology mosaic. Wiley Interdisc Rev Dev Biol 3(1):83–112. doi:10.1002/wdev.121

Kanehisa M, Goto S (2000) KEGG: kyoto encyclopedia of genes and genomes. Nucleic Acids Res 28(1):27–30

Kashou AH, Sharma R, Agarwal A (2013) Assessment of oxidative stress in sperm and semen. Methods Mol Bio (Clifton, NJ) 927:351–361. doi:10.1007/978-1-62703-038-0_30

Kasvandik S, Samuel K, Peters M, Eimre M, Peet N, Roost AM, Padrik L, Paju K, Peil L, Salumets A (2016) Deep quantitative proteomics reveals extensive metabolic reprogramming and cancer-like changes of ectopic endometriotic stromal cells. J Proteome Res 15(2):572–584. doi:10.1021/acs.jproteome.5b00965

Katz-Jaffe MG, Linck DW, Schoolcraft WB, Gardner DK (2005) A proteomic analysis of mammalian preimplantation embryonic development. Reproduction 130(6):899–905. doi:10.1530/rep.1.00854

Katz-Jaffe MG, Gardner DK, Schoolcraft WB (2006) Proteomic analysis of individual human embryos to identify novel biomarkers of development and viability. Fertil Steril 85(1):101–107. doi:10.1016/j.fertnstert.2005.09.011

Katz DF, Morales P, Samuels SJ, Overstreet JW (1990) Mechanisms of filtration of morphologically abnormal human sperm by cervical mucus. Fertil Steril 54(3):513–516

Keshava Prasad TS, Goel R, Kandasamy K, Keerthikumar S, Kumar S, Mathivanan S, Telikicherla D, Raju R, Shafreen B, Venugopal A, Balakrishnan L, Marimuthu A, Banerjee S, Somanathan DS, Sebastian A, Rani S, Ray S, Harrys Kishore CJ, Kanth S, Ahmed M, Kashyap MK, Mohmood R, Ramachandra YL, Krishna V, Rahiman BA, Mohan S, Ranganathan P, Ramabadran S, Chaerkady R, Pandey A (2009) Human protein reference database—2009 update. Nucleic Acids Res 37 (Database issue):D767–D772. doi:10.1093/nar/gkn892

Khan GH, Galazis N, Docheva N, Layfield R, Atiomo W (2015) Overlap of proteomics biomarkers between women with pre-eclampsia and PCOS: a systematic review and biomarker database integration. Hum Reprod 30(1):133–148. doi:10.1093/humrep/deu268

Kim ST, Kim HS, Kim HJ, Kim SG, Kang SY, Lim DB, Kang KY (2003) Prefractionation of protein samples for proteome analysis by sodium dodecyl sulfate-polyacrylamide gel electrophoresis. Mol Cells 16(3):316–322

Kimberly L, Case A, Cheung AP, Sierra S, AlAsiri S, Carranza-Mamane B, Case A, Dwyer C, Graham J, Havelock J, Hemmings R, Lee F, Liu K, Murdock W, Senikas V, Vause TD, Wong BC (2012) Advanced reproductive age and fertility: no. 269, November 2011. Int J Gynaecol Obstet official Organ Int Fed Gynaecol Obstet 117(1):95–102

Kingsmore SF (2006) Multiplexed protein measurement: technologies and applications of protein and antibody arrays. Nat Rev Drug Discov 5(4):310–320. doi:10.1038/nrd2006

Kohler K, Seitz H (2012) Validation processes of protein biomarkers in serum—a cross platform comparison. Sensors (Basel, Switzerland) 12(9):12710–12728. doi:10.3390/s120912710

Korbakis D, Brinc D, Schiza C, Soosaipillai A, Jarvi K, Drabovich AP, Diamandis EP (2015) Immunocapture-selected reaction monitoring screening facilitates the development of ELISA for the measurement of native TEX101 in biological fluids. Mol Cell Proteomics MCP 14 (6):1517–1526. doi:10.1074/mcp.M114.047571

Kouzarides T (2007) Chromatin modifications and their function. Cell 128(4):693–705. doi:10. 1016/j.cell.2007.02.005

Kovac JR, Pastuszak AW, Lamb DJ (2013) The use of genomics, proteomics, and metabolomics in identifying biomarkers of male infertility. Fertil Steril 99(4):998–1007. doi:10.1016/j. fertnstert.2013.01.111

Krejci J, Stixova L, Pagacova E, Legartova S, Kozubek S, Lochmanova G, Zdrahal Z, Sehnalova P, Dabravolski S, Hejatko J, Bartova E (2015) Post-translational Modifications of histones in human sperm. J Cell Biochem 116(10):2195–2209. doi:10.1002/jcb.25170

Kriegel TM, Heidenreich F, Kettner K, Pursche T, Hoflack B, Grunewald S, Poenicke K, Glander HJ, Paasch U (2009) Identification of diabetes- and obesity-associated proteomic changes in human spermatozoa by difference gel electrophoresis. Reprod Biomed Online 19 (5):660–670

Krisher RL, Schoolcraft WB, Katz-Jaffe MG (2015) Omics as a window to view embryo viability. Fertil Steril 103(2):333–341. doi:10.1016/j.fertnstert.2014.12.116

Kroon B, Harrison K, Martin N, Wong B, Yazdani A (2011) Miscarriage karyotype and its relationship with maternal body mass index, age, and mode of conception. Fertil Steril 95 (5):1827–1829. doi:10.1016/j.fertnstert.2010.11.065

Kumar C, Mann M (2009) Bioinformatics analysis of mass spectrometry-based proteomics data sets. FEBS lett 583(11):1703–1712. doi:10.1016/j.febslet.2009.03.035

Kumar V, Hassan MI, Tomar AK, Kashav T, Nautiyal J, Singh S, Singh TP, Yadav S (2009) Proteomic analysis of heparin-binding proteins from human seminal plasma: a step towards identification of molecular markers of male fertility. J Biosci 34(6):899–908

Kupker W, Diedrich K, Edwards RG (1998) Principles of mammalian fertilization. Hum Reprod 13(Suppl 1):20–32

Kyama CM, Mihalyi A, Gevaert O, Waelkens E, Simsa P, Van de Plas R, Meuleman C, De Moor B, D'Hooghe TM (2011) Evaluation of endometrial biomarkers for semi-invasive diagnosis of endometriosis. Fertil Steril 95(4):1338–1343 e1331–1333. doi:10.1016/j.fertnstert. 2010.06.084

Lacerda JI, Del Giudice PT, da Silva BF, Nichi M, Fariello RM, Fraietta R, Restelli AE, Blumer CG, Bertolla RP, Cedenho AP (2011) Adolescent varicocele: improved sperm function after varicocelectomy. Fertil Steril 95(3):994–999. doi:10.1016/j.fertnstert.2010.10.031

Lange V, Picotti P, Domon B, Aebersold R (2008) Selected reaction monitoring for quantitative proteomics: a tutorial. Mol Syst Biol 4:222. doi:10.1038/msb.2008.61

Latham KE, Garrels JI, Chang C, Solter D (1992) Analysis of embryonic mouse development: construction of a high-resolution, two-dimensional gel protein database. Appl Theoret Electrophor Offi J Int Electrophor Soc 2(6):163–170

Lausted C, Lee I, Zhou Y, Qin S, Sung J, Price ND, Hood L, Wang K (2014) Systems approach to neurodegenerative disease biomarker discovery. Ann Rev Pharmacol Toxicol 54:457–481. doi:10.1146/annurev-pharmtox-011613-135928

Law KP, Lim YP (2013) Recent advances in mass spectrometry: data independent analysis and hyper reaction monitoring. Exp Rev Proteomics 10(6):551–566. doi:10.1586/14789450.2013. 858022

Lee DC, Hassan SS, Romero R, Tarca AL, Bhatti G, Gervasi MT, Caruso JA, Stemmer PM, Kim CJ, Hansen LK, Becher N, Uldbjerg N (2011) Protein profiling underscores immuno-logical functions of uterine cervical mucus plug in human pregnancy. J Proteomics 74(6):817–828. doi:10.1016/j.jprot.2011.02.025

Lefievre L, Chen Y, Conner SJ, Scott JL, Publicover SJ, Ford WC, Barratt CL (2007) Human spermatozoa contain multiple targets for protein S-nitrosylation: an alternative mechanism of the modulation of sperm function by nitric oxide? Proteomics 7(17):3066–3084. doi:10.1002/pmic.200700254

Levran D, Farhi J, Nahum H, Glezerman M, Weissman A (2002) Maturation arrest of human oocytes as a cause of infertility: case report. Hum Reprod 17(6):1604–1609

Li LW, Fan LQ, Zhu WB, Nien HC, Sun BL, Luo KL, Liao TT, Tang L, Lu GX (2007) Establishment of a high-resolution 2-D reference map of human spermatozoal proteins from 12 fertile sperm-bank donors. Asian J Androl 9(3):321–329. doi:10.1111/j.1745-7262.2007. 00261.x

Li J, Liu F, Wang H, Liu X, Liu J, Li N, Wan F, Wang W, Zhang C, Jin S, Liu J, Zhu P, Liu Y (2010) Systematic mapping and functional analysis of a family of human epididymal secretory sperm-located proteins. Mol Cell Proteomics MCP 9(11):2517–2528. doi:10.1074/mcp.M110. 001719

Li J, Liu F, Liu X, Liu J, Zhu P, Wan F, Jin S, Wang W, Li N, Liu J, Wang H (2011a) Mapping of the human testicular proteome and its relationship with that of the epididymis and spermatozoa. Mol Cell Proteomics MCP 10(3):M110 004630. doi:10.1074/mcp.M110.004630

Li J, Tan Z, Li M, Xia T, Liu P, Yu W (2011b) Proteomic analysis of endometrium in fertile women during the prereceptive and receptive phases after luteinizing hormone surge. Fertil Steri 95(3):1161–1163. doi:10.1016/j.fertnstert.2010.09.033

Liao TT, Xiang Z, Zhu WB, Fan LQ (2009) Proteome analysis of round-headed and normal spermatozoa by 2-D fluorescence difference gel electrophoresis and mass spectrometry. Asian J Androl 11(6):683–693. doi:10.1038/aja.2009.59

Liu H, Sadygov RG, Yates JR 3rd (2004) A model for random sampling and estimation of relative protein abundance in shotgun proteomics. Anal Chem 76(14):4193–4201. doi:10.1021/ac0498563

Liu AX, Zhu YM, Luo Q, Wu YT, Gao HJ, Zhu XM, Xu CM, Huang HF (2007) Specific peptide patterns of follicular fluids at different growth stages analyzed by matrix-assisted laser desorption/ionization time-of-flight mass spectrometry. Biochimica et biophysica acta 1770 (1):29–38. doi:10.1016/j.bbagen.2006.06.017

Liu Y, Ding J, Reynolds LM, Lohman K, Register TC, De La Fuente A, Howard TD, Hawkins GA, Cui W, Morris J, Smith SG, Barr RG, Kaufman JD, Burke GL, Post W, Shea S, McCall CE, Siscovick D, Jacobs DR Jr, Tracy RP, Herrington DM, Hoeschele I (2013) Methylomics of gene expression in human monocytes. Hum Mol Genet 22(24):5065–5074. doi:10.1093/hmg/ddt356

Liu FJ, Liu X, Han JL, Wang YW, Jin SH, Liu XX, Liu J, Wang WT, Wang WJ (2015) Aged men share the sperm protein PATE1 defect with young asthenozoospermia patients. Hum Reprod 30(4):861–869. doi:10.1093/humrep/dev003

Lo Turco EG, Souza GH, Garcia JS, Ferreira CR, Eberlin MN, Bertolla RP (2010) Effect of endometriosis on the protein expression pattern of follicular fluid from patients submitted to controlled ovarian hyperstimulation for in vitro fertilization. Hum Rep 25(7):1755–1766. doi:10.1093/humrep/deq102

Lo Turco EG, Cordeiro FB, Lopes PH, Gozzo FC, Pilau EJ, Soler TB, da Silva BF, Del Giudice PT, Bertolla RP, Fraietta R, Cedenho AP (2013) Proteomic analysis of follicular fluid from women with and without endometriosis: new therapeutic targets and biomarkers. Mol Reprod Dev 80(6):441–450. doi:10.1002/mrd.22180

Lopez JL (2007) Two-dimensional electrophoresis in proteome expression analysis. J Chromatogr B Anal Technol Biomed Life Sci 849(1–2):190–202. doi:10.1016/j.jchromb. 2006.11.049

Luo KL, Fan LQ, Lu GX (2008) [Proteomic identification of necrozoospermia-related proteins]. Zhonghua nan ke xue =. Natl J Androl 14(5):431–435

MacCoss MJ, Wu CC, Liu H, Sadygov R, Yates JR 3rd (2003) A correlation algorithm for the automated quantitative analysis of shotgun proteomics data. Anal Chem 75(24):6912–6921. doi:10.1021/ac034790h

Machtinger R, Racowsky C (2013) Morphological systems of human embryo assessment and clinical evidence. Reprod Biomed Online 26(3):210–221. doi:10.1016/j.rbmo.2012.10.021

MacLean B, Tomazela DM, Shulman N, Chambers M, Finney GL, Frewen B, Kern R, Tabb DL, Liebler DC, MacCoss MJ (2010) Skyline: an open source document editor for creating and analyzing targeted proteomics experiments. Bioinf (Oxford, England) 26 (7):966-968. doi:10. 1093/bioinformatics/btq054

Maere S, Heymans K, Kuiper M (2005) BiNGO: a Cytoscape plugin to assess overrepresentation of gene ontology categories in biological networks. Bioinformatics (Oxford, England) 21 (16):3448-3449. doi:10.1093/bioinformatics/bti551

Mains LM, Christenson L, Yang B, Sparks AE, Mathur S, Van Voorhis BJ (2011) Identification of apolipoprotein A1 in the human embryonic secretome. Fertil Steril 96(2):422-427.e422. doi:10. 1016/j.fertnstert.2011.05.049

Mallick P, Kuster B (2010) Proteomics: a pragmatic perspective. Nat Biotechnol 28(7):695–709. doi:10.1038/nbt.1658

Mann M, Jensen ON (2003) Proteomic analysis of post-translational modifications. Nat Biotechnol 21(3):255–261. doi:10.1038/nbt0303-255

Manohar M, Khan H, Sirohi VK, Das V, Agarwal A, Pandey A, Siddiqui WA, Dwivedi A (2014) Alteration in endometrial proteins during early- and mid-secretory phases of the cycle in women with unexplained infertility. PloS One 9(11):e111687. doi:10.1371/journal.pone. 0111687

Marcoux J, Cianferani S (2015) Towards integrative structural mass spectrometry: benefits from hybrid approaches. Methods (San Diego, Calif) 89:4–12. doi:10.1016/j.ymeth.2015.05.024

Marianowski P, Szymusik I, Malejczyk J, Hibner M, Wielgos M (2013) Proteomic analysis of eutopic and ectopic endometriotic tissues based on isobaric peptide tags for relative and absolute quantification (iTRAQ) method. Neuro Endocrinol Lett 34(7):717–721

Martinez-Heredia J, de Mateo S, Vidal-Taboada JM, Ballesca JL, Oliva R (2008) Identification of proteomic differences in asthenozoospermic sperm samples. Hum Reprod 23(4):783–791. doi:10.1093/humrep/den024

Martinez-Bartolome S, Deutsch EW, Binz PA, Jones AR, Eisenacher M, Mayer G, Campos A, Canals F, Bech-Serra JJ, Carrascal M, Gay M, Paradela A, Navajas R, Marcilla M, Hernaez ML, Gutierrez-Blazquez MD, Velarde LF, Aloria K, Beaskoetxea J, Medina-Aunon JA, Albar JP (2013) Guidelines for reporting quantitative mass spectrometry based experiments in proteomics. J Proteomics 95:84–88. doi:10.1016/j.jprot.2013.02.026

Martinez-Bartolome S, Binz PA, Albar JP (2014) The Minimal Information about a Proteomics Experiment (MIAPE) from the proteomics standards initiative. Methods Mol Biol (Clifton, NJ) 1072:765–780. doi:10.1007/978-1-62703-631-3_53

Martinez-Heredia J, Estanyol JM, Ballesca JL, Oliva R (2006) Proteomic identification of human sperm proteins. Proteomics 6(15):4356–4369. doi:10.1002/pmic.200600094

Mascarenhas MN, Flaxman SR, Boerma T, Vanderpoel S, Stevens GA (2012) National, regional, and global trends in infertility prevalence since 1990: a systematic analysis of 277 health surveys. PLoS Med 9(12):e1001356. doi:10.1371/journal.pmed.1001356

Massip A, Mulnard J (1980) Time-lapse cinematographic analysis of hatching of normal and frozen-thawed cow blastocysts. J Reprod Fertil 58(2):475–478

Matorras R, Matorras F, Exposito A, Martinez L, Crisol L (2011) Decline in human fertility rates with male age: a consequence of a decrease in male fecundity with aging? Gynecol Obstet Inv 71(4):229–235. doi:10.1159/000319236

Mayeux R (2004) Biomarkers: potential uses and limitations. NeuroRx J Am Soc Exp Neurother 1 (2):182–188. doi:10.1602/neurorx.1.2.182

Mazzocchi F (2012) Complexity and the reductionism-holism debate in systems biology. Wiley Interdisc Rev Syst Biol Med 4(5):413–427. doi:10.1002/wsbm.1181

McReynolds S, Dzieciatkowska M, Stevens J, Hansen KC, Schoolcraft WB, Katz-Jaffe MG (2014) Toward the identification of a subset of unexplained infertility: a sperm proteomic approach. Fertil Steril 102(3):692–699. doi:10.1016/j.fertnstert.2014.05.021

Meng Y, Liu XH, Ma X, Shen Y, Fan L, Leng J, Liu JY, Sha JH (2007) The protein profile of mouse mature cumulus-oocyte complex. Biochimica et biophysica acta 1774(11):1477–1490. doi:10.1016/j.bbapap.2007.08.026

Merelli I, Perez-Sanchez H, Gesing S, D'Agostino D (2014) Managing, analysing, and integrating big data in medical bioinformatics: open problems and future perspectives. Biomed Res Int 2014:134023. doi:10.1155/2014/134023

Mi H, Muruganujan A, Thomas PD (2013) PANTHER in 2013: modeling the evolution of gene function, and other gene attributes, in the context of phylogenetic trees. Nucleic Acids Res 41 (Database issue):D377–D386. doi:10.1093/nar/gks1118

Michalski A, Cox J, Mann M (2011) More than 100,000 detectable peptide species elute in single shotgun proteomics runs but the majority is inaccessible to data-dependent LC-MS/MS. J Proteome Res 10(4):1785–1793. doi:10.1021/pr101060v

Mihm M, Gangooly S, Muttukrishna S (2011) The normal menstrual cycle in women. Anim Reprod Sci 124(3–4):229–236. doi:10.1016/j.anireprosci.2010.08.030

Milardi D, Grande G, Vincenzoni F, Castagnola M, Marana R (2013) Proteomics of human seminal plasma: identification of biomarker candidates for fertility and infertility and the evolution of technology. Mol Reprod Dev 80(5):350–357. doi:10.1002/mrd.22178

Milardi D, Grande G, Vincenzoni F, Giampietro A, Messana I, Castagnola M, Marana R, De Marinis L, Pontecorvi A (2014) Novel biomarkers of androgen deficiency from seminal plasma profiling using high-resolution mass spectrometry. J Clin Endocrinol Metab 99(8):2813–2820. doi:10.1210/jc.2013-4148

Milardi D, Grande G, Vincenzoni F, Messana I, Pontecorvi A, De Marinis L, Castagnola M, Marana R (2012) Proteomic approach in the identification of fertility pattern in seminal plasma of fertile men. Fertil Steril 97(1):67–73 e61. doi:10.1016/j.fertnstert.2011.10.013

Minten MA, Bilby TR, Bruno RG, Allen CC, Madsen CA, Wang Z, Sawyer JE, Tibary A, Neibergs HL, Geary TW, Bauersachs S, Spencer TE (2013) Effects of fertility on gene expression and function of the bovine endometrium. PloS One 8(8):e69444. doi:10.1371/journal.pone.0069444

Mirza SP, Olivier M (2008) Methods and approaches for the comprehensive characterization and quantification of cellular proteomes using mass spectrometry. Physiol Genomics 33(1):3–11. doi:10.1152/physiolgenomics.00292.2007

Mitra A, Richardson RT, O'Rand MG (2010) Analysis of recombinant human semenogelin as an inhibitor of human sperm motility. Biol Reprod 82(3):489–496. doi:10.1095/biolreprod.109.081331

Montag M, Toth B, Strowitzki T (2013) New approaches to embryo selection. Reprod Biomed Online 27(5):539–546. doi:10.1016/j.rbmo.2013.05.013

Mortimer ST, Swan MA (1995) Variable kinematics of capacitating human spermatozoa. Hum Reprod 10(12):3178–3182

Mrazek M, Fulka J Jr (2003) Failure of oocyte maturation: possible mechanisms for oocyte maturation arrest. Hum Reprod 18(11):2249–2252

Munne S, Held KR, Magli CM, Ata B, Wells D, Fragouli E, Baukloh V, Fischer R, Gianaroli L (2012) Intra-age, intercenter, and intercycle differences in chromosome abnormalities in oocytes. Fertil Steril 97(4):935–942. doi:10.1016/j.fertnstert.2012.01.106

Mutsaerts MA, Groen H, Huiting HG, Kuchenbecker WK, Sauer PJ, Land JA, Stolk RP, Hoek A (2012) The influence of maternal and paternal factors on time to pregnancy—a Dutch population-based birth-cohort study: the GECKO Drenthe study. Hum Reprod 27(2):583–593. doi:10.1093/humrep/der429

Naaby-Hansen S (1990) Electrophoretic map of acidic and neutral human spermatozoal proteins. J Reprod Immunol 17(3):167–185

Naaby-Hansen S, Flickinger CJ, Herr JC (1997) Two-dimensional gel electrophoretic analysis of vectorially labeled surface proteins of human spermatozoa. Biol Reprod 56(3):771–787

Nagaraj N, Wisniewski JR, Geiger T, Cox J, Kircher M, Kelso J, Paabo S, Mann M (2011) Deep proteome and transcriptome mapping of a human cancer cell line. Mol Syst Biol 7:548. doi:10.1038/msb.2011.81

Navarrete Santos A, Tonack S, Kirstein M, Kietz S, Fischer B (2004) Two insulin-responsive glucose transporter isoforms and the insulin receptor are developmentally expressed in rabbit preimplantation embryos. Reproduction 128(5):503–516. doi:10.1530/rep.1.00203

Naylor S (2003) Biomarkers: current perspectives and future prospects. Expert Rev Mol Diagn 3 (5):525–529. doi:10.1586/14737159.3.5.525

Nixon B, Mitchell LA, Anderson AL, McLaughlin EA, O'Bryan MK, Aitken RJ (2011) Proteomic and functional analysis of human sperm detergent resistant membranes. J Cell Physiol 226 (10):2651–2665. doi:10.1002/jcp.22615

Noci I, Fuzzi B, Rizzo R, Melchiorri L, Criscuoli L, Dabizzi S, Biagiotti R, Pellegrini S, Menicucci A, Baricordi OR (2005) Embryonic soluble HLA-G as a marker of developmental potential in embryos. Hum Reprod 20(1):138–146. doi:10.1093/humrep/deh572

Nyalwidhe J, Burch T, Bocca S, Cazares L, Green-Mitchell S, Cooke M, Birdsall P, Basu G, Semmes OJ, Oehninger S (2013) The search for biomarkers of human embryo developmental potential in IVF: a comprehensive proteomic approach. Mol Hum Reprod 19(4):250–263. doi:10.1093/molehr/gas063

O'Neill C (2005) The role of paf in embryo physiology. Hum Reprod Update 11(3):215–228. doi:10.1093/humupd/dmi003

Oedit A, Vulto P, Ramautar R, Lindenburg PW, Hankemeier T (2015) Lab-on-a-Chip hyphenation with mass spectrometry: strategies for bioanalytical applications. Curr Opin Biotechnol 31:79–85. doi:10.1016/j.copbio.2014.08.009

Oliva R, Castillo J (2011) Proteomics and the genetics of sperm chromatin condensation. Asian J Androl 13(1):24–30. doi:10.1038/aja.2010.65

Olsen JV, Mann M (2013) Status of large-scale analysis of post-translational modifications by mass spectrometry. Mol Cell Proteomics MCP 12(12):3444–3452. doi:10.1074/mcp.O113.034181

Ong SE, Blagoev B, Kratchmarova I, Kristensen DB, Steen H, Pandey A, Mann M (2002) Stable isotope labeling by amino acids in cell culture, SILAC, as a simple and accurate approach to expression proteomics. Mol Cell Proteomics MCP 1(5):376–386

Ong SE, Foster LJ, Mann M (2003) Mass spectrometric-based approaches in quantitative proteomics. Methods (San Diego, Calif) 29(2):124–130

Orchard S, Ammari M, Aranda B, Breuza L, Briganti L, Broackes-Carter F, Campbell NH, Chavali G, Chen C, del-Toro N, Duesbury M, Dumousseau M, Galeota E, Hinz U, Iannuccelli M, Jagannathan S, Jimenez R, Khadake J, Lagreid A, Licata L, Lovering RC, Meldal B, Melidoni AN, Milagros M, Peluso D, Perfetto L, Porras P, Raghunath A, Ricard-Blum S, Roechert B, Stutz A, Tognolli M, van Roey K, Cesareni G, Hermjakob H (2014) The MIntAct project—IntAct as a common curation platform for 11 molecular interaction databases. Nucleic Acids Res 42(Database issue):D358-D363. doi:10.1093/nar/gkt1115

Orphanou CM, Walton-Williams L, Mountain H, Cassella J (2015) The detection and discrimination of human body fluids using ATR FT-IR spectroscopy. Forensic Sci Int 252: e10–e16. doi:10.1016/j.forsciint.2015.04.020

Ow SY, Salim M, Noirel J, Evans C, Wright PC (2011) Minimising iTRAQ ratio compression through understanding LC-MS elution dependence and high-resolution HILIC fractionation. Proteomics 11(11):2341–2346. doi:10.1002/pmic.201000752

Paasch U, Heidenreich F, Pursche T, Kuhlisch E, Kettner K, Grunewald S, Kratzsch J, Dittmar G, Glander HJ, Hoflack B, Kriegel TM (2011) Identification of increased amounts of eppin protein complex components in sperm cells of diabetic and obese individuals by difference gel electrophoresis. Mol Cell Proteomics MCP 10(8):M110 007187. doi:10.1074/mcp.M110. 007187

Padron OF, Brackett NL, Sharma RK, Lynne CM, Thomas AJ Jr, Agarwal A (1997) Seminal reactive oxygen species and sperm motility and morphology in men with spinal cord injury. Fertil Steril 67(6):1115–1120

Panicker G, Ye Y, Wang D, Unger ER (2010) Characterization of the human cervical mucous proteome. Clin Proteomics 6(1–2):18–28. doi:10.1007/s12014-010-9042-3

Paria BC, Lim H, Das SK, Reese J, Dey SK (2000) Molecular signaling in uterine receptivity for implantation. Semin Cell Dev Biol 11(2):67–76. doi:10.1006/scdb.2000.0153

Parmar T, Sachdeva G, Savardekar L, Katkam RR, Nimbkar-Joshi S, Gadkar-Sable S, Salvi V, Manjramkar DD, Meherji P, Puri CP (2008) Protein repertoire of human uterine fluid during the mid-secretory phase of the menstrual cycle. Hum Reprod 23(2):379–386. doi:10.1093/ humrep/dem367

Parmar T, Gadkar-Sable S, Savardekar L, Katkam R, Dharma S, Meherji P, Puri CP, Sachdeva G (2009) Protein profiling of human endometrial tissues in the midsecretory and proliferative phases of the menstrual cycle. Fertil Steril 92(3):1091–1103. doi:10.1016/j.fertnstert.2008.07. 1734

Parte PP, Rao P, Redij S, Lobo V, D'Souza SJ, Gajbhiye R, Kulkarni V (2012) Sperm phosphoproteome profiling by ultra performance liquid chromatography followed by data independent analysis (LC-MS(E)) reveals altered proteomic signatures in asthenozoospermia. J Proteomics 75(18):5861–5871. doi:10.1016/j.jprot.2012.07.003

Patel S, Ahmed S (2015) Emerging field of metabolomics: big promise for cancer biomarker identification and drug discovery. J Pharm Biomed Anal 107:63–74. doi:10.1016/j.jpba.2014. 12.020

Peng IX, Ogorzalek Loo RR, Shiea J, Loo JA (2008) Reactive-electrospray-assisted laser desorption/ionization for characterization of peptides and proteins. Anal Chem 80(18):6995–7003. doi:10.1021/ac800870c

Pernemalm M, Lehtio J (2013) A novel prefractionation method combining protein and peptide isoelectric focusing in immobilized pH gradient strips. J Proteome Res 12(2):1014–1019. doi:10.1021/pr300817y

Pfeiffer MJ, Siatkowski M, Paudel Y, Balbach ST, Baeumer N, Crosetto N, Drexler HC, Fuellen G, Boiani M (2011) Proteomic analysis of mouse oocytes reveals 28 candidate factors of the "reprogrammome". J Proteome Res 10(5):2140–2153. doi:10.1021/pr100706k

Piehowski PD, Petyuk VA, Orton DJ, Xie F, Moore RJ, Ramirez-Restrepo M, Engel A, Lieberman AP, Albin RL, Camp DG, Smith RD, Myers AJ (2013) Sources of technical variability in quantitative LC-MS proteomics: human brain tissue sample analysis. J Proteome Res 12(5):2128–2137. doi:10.1021/pr301146m

Pilch B, Mann M (2006) Large-scale and high-confidence proteomic analysis of human seminal plasma. Genome biology 7(5):R40. doi:10.1186/gb-2006-7-5-r40

Pilatz A, Lochnit G, Karnati S, Paradowska-Dogan A, Lang T, Schultheiss D, Schuppe HC, Hossain H, Baumgart-Vogt E, Weidner W, Wagenlehner F (2014) Acute epididymitis induces alterations in sperm protein composition. Fertil Steril 101(6):1609–1617 e1601-1605. doi:10. 1016/j.fertnstert.2014.03.011

Pixton KL, Deeks ED, Flesch FM, Moseley FL, Bjorndahl L, Ashton PR, Barratt CL, Brewis IA (2004) Sperm proteome mapping of a patient who experienced failed fertilization at IVF reveals altered expression of at least 20 proteins compared with fertile donors: case report. Human Reprod 19(6):1438–1447. doi:10.1093/humrep/deh224

Plant TM, Zeleznik AJ (2014) Knobil and Neill's physiology of reproduction: two-volume set. Academic Press, Cambridge

Poli M, Ori A, Child T, Jaroudi S, Spath K, Beck M, Wells D (2015) Characterization and quantification of proteins secreted by single human embryos prior to implantation. EMBO Mol Med 7(11):1465–1479. doi:10.15252/emmm.201505344

Primakoff P, Myles DG (2002) Penetration, adhesion, and fusion in mammalian sperm-egg interaction. Science 296(5576):2183–2185. doi:10.1126/science.1072029

Purohit S, Sharma A, She JX (2015) Luminex and other multiplex high throughput technologies for the identification of, and host response to, environmental triggers of type 1 diabetes. BioMed Res Int 2015:326918. doi:10.1155/2015/326918

Rai P, Kota V, Deendayal M, Shivaji S (2010a) Differential proteome profiling of eutopic endometrium from women with endometriosis to understand etiology of endometriosis. J Proteome Res 9(9):4407–4419. doi:10.1021/pr100657s

Rai P, Kota V, Sundaram CS, Deendayal M, Shivaji S (2010b) Proteome of human endometrium: Identification of differentially expressed proteins in proliferative and secretory phase endometrium. Proteomics Clin Appl 4(1):48–59. doi:10.1002/prca.200900094

Regassa A, Rings F, Hoelker M, Cinar U, Tholen E, Looft C, Schellander K, Tesfaye D (2011) Transcriptome dynamics and molecular cross-talk between bovine oocyte and its companion cumulus cells. BMC Genomics 12:57. doi:10.1186/1471-2164-12-57

Regiani T, Cordeiro FB, da Costa Ldo V, Salgueiro J, Cardozo K, Carvalho VM, Perkel KJ, Zylbersztejn DS, Cedenho AP, Lo Turco EG (2015) Follicular fluid alterations in endometriosis: label-free proteomics by MS(E) as a functional tool for endometriosis. Syst Bio Reprod Med 61(5):263–276. doi:10.3109/19396368.2015.1037025

Revelli A, Delle Piane L, Casano S, Molinari E, Massobrio M, Rinaudo P (2009) Follicular fluid content and oocyte quality: from single biochemical markers to metabolomics. Reprod Biol Endocrinol RB&E 7:40. doi:10.1186/1477-7827-7-40

Rifai N, Gillette MA, Carr SA (2006) Protein biomarker discovery and validation: the long and uncertain path to clinical utility. Nat Biotechnol 24. doi:10.1038/nbt1235

Robert M, Gagnon C (1999) Semenogelin I: a coagulum forming, multifunctional seminal vesicle protein. Cell Mol Life Sci (CMLS) 55(6–7):944–960

Rodgers RJ, Irving-Rodgers HF (2010) Formation of the ovarian follicular antrum and follicular fluid. Biol Reprod 82(6):1021–1029. doi:10.1095/biolreprod.109.082941

Rojas J, Chavez-Castillo M, Olivar LC, Calvo M, Mejias J, Rojas M, Morillo J, Bermudez V (2015) Physiologic course of female reproductive function: a molecular look into the prologue of life. J Pregnancy 2015:715735. doi:10.1155/2015/715735

Rolland AD, Lavigne R, Dauly C, Calvel P, Kervarrec C, Freour T, Evrard B, Rioux-Leclercq N, Auger J, Pineau C (2013) Identification of genital tract markers in the human seminal plasma using an integrative genomics approach. Hum Reprod 28(1):199–209. doi:10.1093/humrep/des360

Ross CA, Poirier MA (2004) Protein aggregation and neurodegenerative disease. Nat Med 10 (Suppl):S10–S17. doi:10.1038/nm1066

Ross PL, Huang YN, Marchese JN, Williamson B, Parker K, Hattan S, Khainovski N, Pillai S, Dey S, Daniels S, Purkayastha S, Juhasz P, Martin S, Bartlet-Jones M, He F, Jacobson A, Pappin DJ (2004) Multiplexed protein quantitation in Saccharomyces cerevisiae using amine-reactive isobaric tagging reagents. Mol Cell Proteomics MCP 3(12):1154–1169. doi:10.1074/mcp.M400129-MCP200

Rouillard AD, Wang Z, Ma'ayan A (2015) Reprint of "Abstraction for data integration: fusing mammalian molecular, cellular and phenotype big datasets for better knowledge extraction". Comput Biol Chem 59 Pt B:123–138. doi:10.1016/j.compbiolchem.2015.08.005

Said T, Agarwal A, Grunewald S, Rasch M, Baumann T, Kriegel C, Li L, Glander HJ, Thomas AJ Jr, Paasch U (2006) Selection of nonapoptotic spermatozoa as a new tool for enhancing assisted reproduction outcomes: an in vitro model. Biol Reprod 74(3):530–537. doi:10.1095/biolreprod.105.046607

Sakkas D, Lu C, Zulfikaroglu E, Neuber E, Taylor HS (2003) A soluble molecule secreted by human blastocysts modulates regulation of HOXA10 expression in an epithelial endometrial cell line. Fertil Steril 80(5):1169–1174

Samanta L, Swain N, Ayaz A, Venugopal V, Agarwal A (1860) Post-Translational Modifications in sperm Proteome: The Chemistry of Proteome diversifications in the Pathophysiology of male factor infertility. Biochimica et biophysica acta 1860(7):1450–1465. doi:10.1016/j.bbagen.2016.04.001

Santagata S, Eberlin LS, Norton I, Calligaris D, Feldman DR, Ide JL, Liu X, Wiley JS, Vestal ML, Ramkissoon SH, Orringer DA, Gill KK, Dunn IF, Dias-Santagata D, Ligon KL, Jolesz FA, Golby AJ, Cooks RG, Agar NY (2014) Intraoperative mass spectrometry mapping of an onco-metabolite to guide brain tumor surgery. Proceedings of the National Academy of Sciences of the United States of America 111(30):11121–11126. doi:10.1073/pnas.1404724111

Santonocito M, Vento M, Guglielmino MR, Battaglia R, Wahlgren J, Ragusa M, Barbagallo D, Borzi P, Rizzari S, Maugeri M, Scollo P, Tatone C, Valadi H, Purrello M, Di Pietro C (2014) Molecular characterization of exosomes and their microRNA cargo in human follicular fluid: bioinformatic analysis reveals that exosomal microRNAs control pathways involved in follicular maturation. Fertil Steril 102(6):1751–1761.e1751. doi:10.1016/j.fertnstert.2014.08.005

Sargent I, Swales A, Ledee N, Kozma N, Tabiasco J, Le Bouteiller P (2007) sHLA-G production by human IVF embryos: can it be measured reliably? J Reprod Immunol 75(2):128–132. doi:10.1016/j.jri.2007.03.005

Savaryn JP, Catherman AD, Thomas PM, Abecassis MM, Kelleher NL (2013) The emergence of top-down proteomics in clinical research. Genome Med 5(6):53. doi:10.1186/gm457

Scherl A (2015) Clinical protein mass spectrometry. Methods (San Diego, Calif) 81:3-14. doi:10.1016/j.ymeth.2015.02.015

Schiller J, Arnhold J, Glander HJ, Arnold K (2000) Lipid analysis of human spermatozoa and seminal plasma by MALDI-TOF mass spectrometry and NMR spectroscopy—effects of freezing and thawing. Chem Phys Lipids 106(2):145–156

Schweigert FJ, Gericke B, Wolfram W, Kaisers U, Dudenhausen JW (2006) Peptide and protein profiles in serum and follicular fluid of women undergoing IVF. Hum Reprod 21(11):2960–2968. doi:10.1093/humrep/del257

Scotchie JG, Fritz MA, Mocanu M, Lessey BA, Young SL (2009) Proteomic analysis of the luteal endometrial secretome. Reprod Sci 16(9):883–893. doi:10.1177/1933719109337165

Secciani F, Bianchi L, Ermini L, Cianti R, Armini A, La Sala GB, Focarelli R, Bini L, Rosati F (2009) Protein profile of capacitated versus ejaculated human sperm. J Proteome Res 8 (7):3377–3389. doi:10.1021/pr900031r

Seli E, Sakkas D, Scott R, Kwok SC, Rosendahl SM, Burns DH (2007) Noninvasive metabolomic profiling of embryo culture media using Raman and near-infrared spectroscopy correlates with reproductive potential of embryos in women undergoing in vitro fertilization. Fertil Steril 88 (5):1350–1357. doi:10.1016/j.fertnstert.2007.07.1390

Seli E, Babayev E, Collins SC, Nemeth G, Horvath TL (2014) Minireview: Metabolism of female reproduction: regulatory mechanisms and clinical implications. Mol Endocrinol 28(6):790–804. doi:10.1210/me.2013-1413

Seo J, Lee KJ (2004) Post-translational modifications and their biological functions: proteomic analysis and systematic approaches. J Biochem Mol Biol 37(1):35–44

Sethuraman M, McComb ME, Heibeck T, Costello CE, Cohen RA (2004) Isotope-coded affinity tag approach to identify and quantify oxidant-sensitive protein thiols. Mol Cell Proteomics MCP 3(3):273–278. doi:10.1074/mcp.T300011-MCP200

Sharma R, Agarwal A, Mohanty G, Du Plessis SS, Gopalan B, Willard B, Yadav SP, Sabanegh E (2013a) Proteomic analysis of seminal fluid from men exhibiting oxidative stress. Reprod Biol Endocrinol (RB&E) 11:85. doi:10.1186/1477-7827-11-85

Sharma R, Agarwal A, Mohanty G, Hamada AJ, Gopalan B, Willard B, Yadav S, du Plessis S (2013b) Proteomic analysis of human spermatozoa proteins with oxidative stress. Reprod Biol Endocrinol (RB&E) 11:48. doi:10.1186/1477-7827-11-48

Sharma R, Agarwal A, Mohanty G, Jesudasan R, Gopalan B, Willard B, Yadav SP, Sabanegh E (2013c) Functional proteomic analysis of seminal plasma proteins in men with various semen parameters. Reprod Biol Endocrinol (RB&E) 11:38. doi:10.1186/1477-7827-11-38

Sharma R, Biedenharn KR, Fedor JM, Agarwal A (2013d) Lifestyle factors and reproductive health: taking control of your fertility. Reprod Biol Endocrinol (RB&E) 11:66. doi:10.1186/1477-7827-11-66

Sharma R, Ahmad G, Esteves SC, Agarwal A (2016) Terminal deoxynucleotidyl transferase dUTP nick end labeling (TUNEL) assay using bench top flow cytometer for evaluation of sperm DNA fragmentation in fertility laboratories: protocol, reference values, and quality control. J Assist Reprod Genet 33(2):291–300. doi:10.1007/s10815-015-0635-7

Shaw JL, Smith CR, Diamandis EP (2007) Proteomic analysis of human cervico-vaginal fluid. J Proteome Res 6(7):2859–2865. doi:10.1021/pr0701658

Sher G, Keskintepe L, Fisch JD, Acacio BA, Ahlering P, Batzofin J, Ginsburg M (2005) Soluble human leukocyte antigen G expression in phase I culture media at 46 hours after fertilization predicts pregnancy and implantation from day 3 embryo transfer. Fertil Steri 83(5):1410–1413. doi:10.1016/j.fertnstert.2004.11.061

Shi CZ, Collins HW, Garside WT, Buettger CW, Matschinsky FM, Heyner S (1994) Protein databases for compacted eight-cell and blastocyst-stage mouse embryos. Mol Reprod Dev 37 (1):34–47. doi:10.1002/mrd.1080370106

Siva AB, Kameshwari DB, Singh V, Pavani K, Sundaram CS, Rangaraj N, Deenadayal M, Shivaji S (2010) Proteomics-based study on asthenozoospermia: differential expression of proteasome alpha complex. Mol Hum Reprod 16(7):452–462. doi:10.1093/molehr/gaq009

Smith R, Kaune H, Parodi D, Madariaga M, Morales I, Rios R, Castro A (2007) Extent of sperm DNA damage in spermatozoa from men examined for infertility. Relationship with oxidative stress. Revista medica de Chile 135(3):279–286. doi:10.4067/S0034-98872007000300001

Sobrero AJ, Macleod J (1962) The immediate postcoital test. Fertil Steril 13:184–189

Spencer TE, Forde N, Dorniak P, Hansen TR, Romero JJ, Lonergan P (2013) Conceptus-derived prostaglandins regulate gene expression in the endometrium prior to pregnancy recognition in ruminants. Reproduction 146(4):377–387. doi:10.1530/rep-13-0165

Spitzer D, Murach KF, Lottspeich F, Staudach A, Illmensee K (1996) Different protein patterns derived from follicular fluid of mature and immature human follicles. Hum Reprod 11(4):798–807

Starita-Geribaldi M, Poggioli S, Zucchini M, Garin J, Chevallier D, Fenichel P, Pointis G (2001) Mapping of seminal plasma proteins by two-dimensional gel electrophoresis in men with normal and impaired spermatogenesis. Mol Hum Reprod 7(8):715–722

Starita-Geribaldi M, Roux F, Garin J, Chevallier D, Fenichel P, Pointis G (2003) Development of narrow immobilized pH gradients covering one pH unit for human seminal plasma proteomic analysis. Proteomics 3(8):1611–1619. doi:10.1002/pmic.200300493

Stark C, Breitkreutz BJ, Reguly T, Boucher L, Breitkreutz A, Tyers M (2006) BioGRID: a general repository for interaction datasets. Nucleic Acids Res 34(Database issue):D535–D539. doi:10.1093/nar/gkj109

Stevanato J, Bertolla RP, Barradas V, Spaine DM, Cedenho AP, Ortiz V (2008) Semen processing by density gradient centrifugation does not improve sperm apoptotic deoxyribonucleic acid fragmentation rates. Fertil Steril 90(3):889–890. doi:10.1016/j.fertnstert.2007.01.059

Stewart AF, Kim ED (2011) Fertility concerns for the aging male. Urology 78(3):496–499. doi:10.1016/j.urology.2011.06.010

Stricker R, Eberhart R, Chevailler MC, Quinn FA, Bischof P, Stricker R (2006) Establishment of detailed reference values for luteinizing hormone, follicle stimulating hormone, estradiol, and progesterone during different phases of the menstrual cycle on the Abbott ARCHITECT analyzer. Clin Chem Lab Med 44(7):883–887. doi:10.1515/CCLM.2006.160

Strimbu K, Tavel JA (2010) What are biomarkers? Curr Opin HIV AIDS 5(6):463–466. doi:10. 1097/COH.0b013e32833ed177

Suarez SS (2008) Control of hyperactivation in sperm. Hum Reprod Update 14(6):647–657. doi:10.1093/humupd/dmn029

Suarez SS (2016) Mammalian sperm interactions with the female reproductive tract. Cell Tissue Res 363(1):185–194. doi:10.1007/s00441-015-2244-2

Suarez SS, Pacey AA (2006) Sperm transport in the female reproductive tract. Hum Reprod Update 12(1):23–37. doi:10.1093/humupd/dmi047

Sun G, Jiang M, Zhou T, Guo Y, Cui Y, Guo X, Sha J (2014) Insights into the lysine acetylproteome of human sperm. J Proteomics 109:199–211. doi:10.1016/j.jprot.2014.07.002

Surinova S, Schiess R, Huttenhain R, Cerciello F, Wollscheid B, Aebersold R (2011) On the development of plasma protein biomarkers. J Proteome Res 10(1):5–16. doi:10.1021/ pr1008515

Szklarczyk D, Franceschini A, Wyder S, Forslund K, Heller D, Huerta-Cepas J, Simonovic M, Roth A, Santos A, Tsafou KP, Kuhn M, Bork P, Jensen LJ, von Mering C (2015) STRING v10: protein-protein interaction networks, integrated over the tree of life. Nucleic Acids Res 43 (Database issue):D447–D452. doi:10.1093/nar/gku1003

Tabb DL, McDonald WH, Yates JR 3rd (2002) DTASelect and contrast: tools for assembling and comparing protein identifications from shotgun proteomics. J Proteome Res 1(1):21–26

Tahmasbpour E, Balasubramanian D, Agarwal A (2014) A multi-faceted approach to understanding male infertility: gene mutations, molecular defects and assisted reproductive techniques (ART). J Assist Reprod Genet 31(9):1115–1137. doi:10.1007/s10815-014-0280-6

Tanaka K, Waki H, Ido Y, Akita S, Yoshida Y, Yoshida T, Matsuo T (1988) Protein and polymer analyses up to m/z 100 000 by laser ionization time-of-flight mass spectrometry. Rapid Commun Mass Spectrom 2(8):151–153

Tanca A, Biosa G, Pagnozzi D, Addis MF, Uzzau S (2013) Comparison of detergent-based sample preparation workflows for LTQ-Orbitrap analysis of the Escherichia coli proteome. Proteomics 13(17):2597–2607. doi:10.1002/pmic.201200478

Tang LJ, De Seta F, Odreman F, Venge P, Piva C, Guaschino S, Garcia RC (2007) Proteomic analysis of human cervical-vaginal fluids. J Proteome Res 6(7):2874–2883. doi:10.1021/ pr0700899

Taylor SC, Posch A (2014) The design of a quantitative western blot experiment. BioMed Res Int 2014:361590. doi:10.1155/2014/361590

Thacker S, Yadav SP, Sharma RK, Kashou A, Willard B, Zhang D, Agarwal A (2011) Evaluation of sperm proteins in infertile men: a proteomic approach. Fertil Steril 95(8):2745–2748. doi:10. 1016/j.fertnstert.2011.03.112

Thimon V, Frenette G, Saez F, Thabet M, Sullivan R (2008) Protein composition of human epididymosomes collected during surgical vasectomy reversal: a proteomic and genomic approach. Hum Reprod 23(8):1698–1707. doi:10.1093/humrep/den181

Thurin A, Hausken J, Hillensjo T, Jablonowska B, Pinborg A, Strandell A, Bergh C (2004) Elective single-embryo transfer versus double-embryo transfer in in vitro fertilization. N Engl J Med 351(23):2392–2402. doi:10.1056/NEJMoa041032

Tokushige N, Markham R, Crossett B, Ahn SB, Nelaturi VL, Khan A, Fraser IS (2011) Discovery of a novel biomarker in the urine in women with endometriosis. Fertil Steril 95(1):46–49. doi:10.1016/j.fertnstert.2010.05.016

Tomar AK, Sooch BS, Raj I, Singh S, Yadav S (2013) Interaction analysis identifies semenogelin I fragments as new binding partners of PIP in human seminal plasma. Int J Biol Macromol 52:296–299. doi:10.1016/j.ijbiomac.2012.10.011

Tran JC, Zamdborg L, Ahlf DR, Lee JE, Catherman AD, Durbin KR, Tipton JD, Vellaichamy A, Kellie JF, Li M, Wu C, Sweet SM, Early BP, Siuti N, LeDuc RD, Compton PD, Thomas PM, Kelleher NL (2011) Mapping intact protein isoforms in discovery mode using top-down proteomics. Nature 480(7376):254–258. doi:10.1038/nature10575

Tsai CF, Wang YT, Chen YR, Lai CY, Lin PY, Pan KT, Chen JY, Khoo KH, Chen YJ (2008) Immobilized metal affinity chromatography revisited: pH/acid control toward high selectivity in phosphoproteomics. J Proteome Res 7(9):4058–4069. doi:10.1021/pr800364d

Twigt J, Steegers-Theunissen RP, Bezstarosti K, Demmers JA (2012) Proteomic analysis of the microenvironment of developing oocytes. Proteomics 12(9):1463–1471. doi:10.1002/pmic. 201100240

UniProt C KLK3 - Prostate-specific antigen precursor - Homo sapiens (Human) (2016) http:// www.uniprot.org/uniprot/P07288. Accessed 14 Sept 2016

UniProt C (2015) UniProt: a hub for protein information. Nucleic acids research 43 Database issue):D204–D212. doi:10.1093/nar/gku989

Upadhyay RD, Balasinor NH, Kumar AV, Sachdeva G, Parte P (1834) Dumasia K (2013) Proteomics in reproductive biology: beacon for unraveling the molecular complexities. Biochimica et biophysica acta 1:8–15. doi:10.1016/j.bbapap.2012.10.004

Vandermarliere E, Mueller M, Martens L (2013) Getting intimate with trypsin, the leading protease in proteomics. Mass Spectrom Rev 32(6):453–465. doi:10.1002/mas.21376

Varshini J, Srinag BS, Kalthur G, Krishnamurthy H, Kumar P, Rao SB, Adiga SK (2012) Poor sperm quality and advancing age are associated with increased sperm DNA damage in infertile men. Andrologia 44(Suppl 1):642–649. doi:10.1111/j.1439-0272.2011.01243.x

Vasen G, Battistone MA, Croci DO, Brukman NG, Weigel Munoz M, Stupirski JC, Rabinovich GA, Cuasnicu PS (2015) The galectin-1-glycan axis controls sperm fertilizing capacity by regulating sperm motility and membrane hyperpolarization. FASEB J Official Publ Fed Am Soc Exp Biol 29(10):4189–4200. doi:10.1096/fj.15-270975

Vercammen MJ, Verloes A, Van de Velde H, Haentjens P (2008) Accuracy of soluble human leukocyte antigen-G for predicting pregnancy among women undergoing infertility treatment: meta-analysis. Hum Reprod Update 14(3):209–218. doi:10.1093/humupd/dmn007

Verrills NM (2006) Clinical proteomics: present and future prospects. Clin Biochem Rev/Aust Assoc Clin Biochemists 27(2):99–116

Vigodner M, Shrivastava V, Gutstein LE, Schneider J, Nieves E, Goldstein M, Feliciano M, Callaway M (2013) Localization and identification of sumoylated proteins in human sperm: excessive sumoylation is a marker of defective spermatozoa. Hum Reprod 28(1):210–223. doi:10.1093/humrep/des317

von Haller PD, Yi E, Donohoe S, Vaughn K, Keller A, Nesvizhskii AI, Eng J, Li XJ, Goodlett DR, Aebersold R, Watts JD (2003) The application of new software tools to quantitative protein profiling via isotope-coded affinity tag (ICAT) and tandem mass spectrometry: I. Statistically annotated datasets for peptide sequences and proteins identified via the application of ICAT and tandem mass spectrometry to proteins copurifying with T cell lipid rafts. Mol Cell Proteomics (MCP) 2(7):426–427. doi:10.1074/mcp.D300002-MCP200

Wang X, Chen C, Wang L, Chen D, Guang W, French J (2003) Conception, early pregnancy loss, and time to clinical pregnancy: a population-based prospective study. Fertil Steril 79(3): 577–584

Wang Y, Puscheck EE, Lewis JJ, Trostinskaia AB, Wang F, Rappolee DA (2005) Increases in phosphorylation of SAPK/JNK and p38MAPK correlate negatively with mouse embryo development after culture in different media. Fertil Steril 83(Suppl 1):1144–1154. doi:10.1016/ j.fertnstert.2004.08.038

Wang J, Wang J, Zhang HR, Shi HJ, Ma D, Zhao HX, Lin B, Li RS (2009) Proteomic analysis of seminal plasma from asthenozoospermia patients reveals proteins that affect oxidative stress responses and semen quality. Asian J Androl 11(4):484–491. doi:10.1038/aja.2009.26

Wang G, Guo Y, Zhou T, Shi X, Yu J, Yang Y, Wu Y, Wang J, Liu M, Chen X, Tu W, Zeng Y, Jiang M, Li S, Zhang P, Zhou Q, Zheng B, Yu C, Zhou Z, Guo X, Sha J (2013a) In-depth proteomic analysis of the human sperm reveals complex protein compositions. J Proteomics 79:114–122. doi:10.1016/j.jprot.2012.12.008

Wang G, Wu Y, Zhou T, Guo Y, Zheng B, Wang J, Bi Y, Liu F, Zhou Z, Guo X, Sha J (2013b) Mapping of the N-linked glycoproteome of human spermatozoa. J Proteome Res 12(12):5750–5759. doi:10.1021/pr400753f

Wang J, Qi L, Huang S, Zhou T, Guo Y, Wang G, Guo X, Zhou Z, Sha J (2015) Quantitative phosphoproteomics analysis reveals a key role of insulin growth factor 1 receptor (IGF1R) tyrosine kinase in human sperm capacitation. Mol Cell Proteomics (MCP) 14(4):1104–1112. doi:10.1074/mcp.M114.045468

Wang S, Kou Z, Jing Z, Zhang Y, Guo X, Dong M, Wilmut I, Gao S (2010) Proteome of mouse oocytes at different developmental stages. Proc Natl Acad Sci U.S.A 107(41):17639–17644. doi:10.1073/pnas.1013185107

Wang K, Huang C, Nice E (2014a) Recent advances in proteomics: towards the human proteome. Biomed Chromatogr (BMC) 28(6):848–857. doi:10.1002/bmc.3157

Wang L, Liu HY, Shi HH, Lang JH, Sun W (2014b) Urine peptide patterns for non-invasive diagnosis of endometriosis: a preliminary prospective study. Eur J Obstet Gynecol Reprod Biol 177:23–28. doi:10.1016/j.ejogrb.2014.03.011

Warner CM, Lampton PW, Newmark JA, Cohen J (2008) Symposium: innovative techniques in human embryo viability assessment. Soluble human leukocyte antigen-G and pregnancy success. Reprod Biomed Online 17(4):470–485

Washburn MP, Wolters D, Yates JR 3rd (2001) Large-scale analysis of the yeast proteome by multidimensional protein identification technology. Nat Biotechnol 19(3):242–247. doi:10.1038/85686

Weckwerth W (2011) Green systems biology—from single genomes, proteomes and metabolomes to ecosystems research and biotechnology. J Proteomics 75(1):284–305. doi:10.1016/j.jprot.2011.07.010

Westermeier R (2014) Looking at proteins from two dimensions: a review on five decades of 2D electrophoresis. Arch Physiol Biochem 120(5):168–172. doi:10.3109/13813455.2014.945188

Weston LA, Bauer KM, Hummon AB (2013) Comparison of bottom-up proteomic approaches for LC-MS analysis of complex proteomes. Anal Methods 5(18). doi:10.1039/C3AY40853A

Whitehouse CM, Dreyer RN, Yamashita M, Fenn JB (1985) Electrospray interface for liquid chromatographs and mass spectrometers. Anal Chem 57(3):675–679

Wilcox AJ, Weinberg CR, Baird DD (1995) Timing of sexual intercourse in relation to ovulation. Effects on the probability of conception, survival of the pregnancy, and sex of the baby. N Engl J Med 333(23):1517–1521. doi:10.1056/NEJM199512073332301

Wilkins MR, Pasquali C, Appel RD, Ou K, Golaz O, Sanchez JC, Yan JX, Gooley AA, Hughes G, Humphery-Smith I, Williams KL, Hochstrasser DF (1996) From proteins to proteomes: large scale protein identification by two-dimensional electrophoresis and amino acid analysis. Biotechnology (N Y) 14(1):61–65

Williams EG, Auwerx J (2015) The convergence of systems and reductionist approaches in complex trait analysis. Cell 162(1):23–32. doi:10.1016/j.cell.2015.06.024

Williamson AJ, Smith DL, Blinco D, Unwin RD, Pearson S, Wilson C, Miller C, Lancashire L, Lacaud G, Kouskoff V, Whetton AD (2008) Quantitative proteomics analysis demonstrates post-transcriptional regulation of embryonic stem cell differentiation to hematopoiesis. Mol Cell Proteomics (MCP) 7(3):459–472. doi:10.1074/mcp.M700370-MCP200

Wisniewski JR, Zougman A, Nagaraj N, Mann M (2009) Universal sample preparation method for proteome analysis. Nat Methods 6(5):359–362. doi:10.1038/nmeth.1322

Wolfler MM, Meinhold-Heerlein IM, Sohngen L, Rath W, Knuchel R, Neulen J, Maass N, Henkel C (2011) Two-dimensional gel electrophoresis in peritoneal fluid samples identifies differential protein regulation in patients suffering from peritoneal or ovarian endometriosis. Fertil Steril 95(8):2764–2768. doi:10.1016/j.fertnstert.2011.03.061

World Health Organization DoRHaR (2010) WHO laboratory manual for the examination and processing of human semen. World Health Organ

Xia J, Mandal R, Sinelnikov IV, Broadhurst D, Wishart DS (2012) MetaboAnalyst 2.0—a comprehensive server for metabolomic data analysis. Nucleic Acids Res 40(Web Server issue): W127–W133. doi:10.1093/nar/gks374

Xu W, Hu H, Wang Z, Chen X, Yang F, Zhu Z, Fang P, Dai J, Wang L, Shi H, Li Z, Qiao Z (2012) Proteomic characteristics of spermatozoa in normozoospermic patients with infertility. J Proteomics 75(17):5426–5436. doi:10.1016/j.jprot.2012.06.021

Xu HM, Deng HT, Liu CD, Chen YL, Zhang ZY (2015) Phosphoproteomics Analysis of Endometrium in Women with or without Endometriosis. Chin Med J 128(19):2617–2624. doi:10.4103/0366-6999.166022

Yamakawa K, Yoshida K, Nishikawa H, Kato T, Iwamoto T (2007) Comparative analysis of interindividual variations in the seminal plasma proteome of fertile men with identification of potential markers for azoospermia in infertile patients. J Androl 28(6):858–865. doi:10.2164/jandrol.107.002824

Yan Q (2014) Translational bioinformatics approaches for systems and dynamical medicine. Methods Mol Biol (Clifton, NJ) 1175:19–34. doi:10.1007/978-1-4939-0956-8_2

Yan W, Lee H, Yi EC, Reiss D, Shannon P, Kwieciszewski BK, Coito C, Li XJ, Keller A, Eng J, Galitski T, Goodlett DR, Aebersold R, Katze MG (2004) System-based proteomic analysis of the interferon response in human liver cells. Genome Biol 5(8):R54. doi:10.1186/gb-2004-5-8-r54

Yang H, Zhou B, Prinz M, Siegel D (2012) Proteomic analysis of menstrual blood. Mol Cell Proteomics(MCP) 11(10):1024–1035. doi:10.1074/mcp.M112.018390

Yang H, Zhou B, Deng H, Prinz M, Siegel D (2013) Body fluid identification by mass spectrometry. Int J Legal Med 127(6):1065–1077. doi:10.1007/s00414-013-0848-1

Yang X, Liu F, Yan Y, Zhou T, Guo Y, Sun G, Zhou Z, Zhang W, Guo X, Sha J (2015) Proteomic analysis of N-glycosylation of human seminal plasma. Proteomics 15(7):1255–1258. doi:10.1002/pmic.201400203

Yao YQ, Barlow DH, Sargent IL (2005) Differential expression of alternatively spliced transcripts of HLA-G in human preimplantation embryos and inner cell masses. J Immunol (Baltimore, Md: 1950) 175(12):8379–8385

Yates JR, Ruse CI, Nakorchevsky A (2009) Proteomics by mass spectrometry: approaches, advances, and applications. Ann Rev Biomed Eng 11:49–79. doi:10.1146/annurev-bioeng-061008-124934

Yu H, Diao H, Wang C, Lin Y, Yu F, Lu H, Xu W, Li Z, Shi H, Zhao S, Zhou Y, Zhang Y (2015) Acetylproteomic analysis reveals functional implications of lysine acetylation in human spermatozoa (sperm). Mol Cell Proteomics (MCP) 14(4):1009–1023. doi:10.1074/mcp.M114.041384

Yurttas P, Morency E, Coonrod SA (2010) Use of proteomics to identify highly abundant maternal factors that drive the egg-to-embryo transition. Reproduction 139(5):809–823. doi:10.1530/REP-09-0538

Zaneveld LJ, De Jonge CJ, Anderson RA, Mack SR (1991) Human sperm capacitation and the acrosome reaction. Hum Reprod 6(9):1265–1274

Zangbar MS, Keshtgar S, Zolghadri J, Gharesi-Fard B (2016) Antisperm protein targets in azoospermia men. J Hum Reprod Sci 9(1):47–52. doi:10.4103/0974-1208.178629

Zegels G, Van Raemdonck GA, Coen EP, Tjalma WA, Van Ostade XW (2009) Comprehensive proteomic analysis of human cervical-vaginal fluid using colposcopy samples. Proteome science 7:17. doi:10.1186/1477-5956-7-17

Zegers-Hochschild F, Adamson GD, de Mouzon J, Ishihara O, Mansour R, Nygren K, Sullivan E (2009) Vanderpoel S (2009) International Committee for Monitoring Assisted Reproductive Technology (ICMART) and the World Health Organization (WHO) revised glossary of ART terminology. Fertil Steril 92(5):1520–1524. doi:10.1016/j.fertnstert.2009.09.009

Zhao SY, Qiao J, Li MZ, Zhang XW, Yu JK, Li R (2008) Preliminary study of protein expression profiling of PCOS on different state. Zhonghua yi xue za zhi 88(1):7–11

Zhang P, Ni X, Guo Y, Guo X, Wang Y, Zhou Z, Huo R, Sha J (2009) Proteomic-based identification of maternal proteins in mature mouse oocytes. BMC Genomics 10:348. doi:10. 1186/1471-2164-10-348

Zhang Y, Fonslow BR, Shan B, Baek MC, Yates JR 3rd (2013) Protein analysis by shotgun/bottom-up proteomics. Chem Rev 113(4):2343–2394. doi:10.1021/cr3003533

Zhao Y, Jensen ON (2009) Modification-specific proteomics: strategies for characterization of post-translational modifications using enrichment techniques. Proteomics 9(20):4632–4641. doi:10.1002/pmic.200900398

Zhao YY, Wu SP, Liu S, Zhang Y, Lin RC (2014) Ultra-performance liquid chromatography-mass spectrometry as a sensitive and powerful technology in lipidomic applications. Chemico-Biological Interact 220:181–192. doi:10.1016/j.cbi.2014.06.029

Zheng J, Li N, Ridyard M, Dai H, Robbins SM, Li L (2005) Simple and robust two-layer matrix/sample preparation method for MALDI MS/MS analysis of peptides. J Proteome Res 4 (5):1709–1716. doi:10.1021/pr050157w

Zhou S, Yi T, Liu R, Bian C, Qi X, He X, Wang K, Li J, Zhao X, Huang C, Wei Y (2012) Proteomics identification of annexin A2 as a key mediator in the metastasis and proangio-genesis of endometrial cells in human adenomyosis. Mol Cell Proteomics (MCP) 11(7):M112 017988. doi:10.1074/mcp.M112.017988

Zhu Y, Wu Y, Jin K, Lu H, Liu F, Guo Y, Yan F, Shi W, Liu Y, Cao X, Hu H, Zhu H, Guo X, Sha J, Li Z, Zhou Z (2013) Differential proteomic profiling in human spermatozoa that did or did not result in pregnancy via IVF and AID. Proteomics Clin Appl 7(11–12):850–858. doi:10. 1002/prca.201200078

Ziebe S, Petersen K, Lindenberg S, Andersen AG, Gabrielsen A, Andersen AN (1997) Embryo morphology or cleavage stage: how to select the best embryos for transfer after in-vitro fertilization. Hum Reprod 12(7):1545–1549

Zini A, Mak V, Phang D, Jarvi K (1999) Potential adverse effect of semen processing on human sperm deoxyribonucleic acid integrity. Fertil Steril 72(3):496–499

Zini A, Finelli A, Phang D, Jarvi K (2000) Influence of semen processing technique on human sperm DNA integrity. Urology 56(6):1081–1084

Zuo X, Speicher DW (2002) Comprehensive analysis of complex proteomes using microscale solution isoelectrofocusing prior to narrow pH range two-dimensional electrophoresis. Proteomics 2(1):58–68

Zylbersztejn DS, Andreoni C, Del Giudice PT, Spaine DM, Borsari L, Souza GH, Bertolla RP, Fraietta R (2013) Proteomic analysis of seminal plasma in adolescents with and without varicocele. Fertil Steril 99(1):92–98. doi:10.1016/j.fertnstert.2012.08.048